Lecture Notes in Computer Science 15184

Founding Editors

Gerhard Goos
Juris Hartmanis

The series Lecture Notes in Computer Science (LNCS), including its subseries Lecture Notes in Artificial Intelligence (LNAI) and Lecture Notes in Bioinformatics (LNBI), has established itself as a medium for the publication of new developments in computer science and information technology research, teaching, and education.

LNCS enjoys close cooperation with the computer science R & D community, the series counts many renowned academics among its volume editors and paper authors, and collaborates with prestigious societies. Its mission is to serve this international community by providing an invaluable service, mainly focused on the publication of conference and workshop proceedings and postproceedings. LNCS commenced publication in 1973.

Zhongying Deng · Yiqing Shen ·
Hyunwoo J. Kim · Won-Ki Jeong ·
Angelica I. Aviles-Rivero · Junjun He ·
Shaoting Zhang
Editors

Foundation Models for General Medical AI

Second International Workshop, MedAGI 2024
Held in Conjunction with MICCAI 2024
Marrakesh, Morocco, October 6, 2024
Proceedings

 Springer

Editors
Zhongying Deng (ID)
University of Cambridge
Cambridge, UK

Yiqing Shen (ID)
Johns Hopkins University
Baltimore, MD, USA

Hyunwoo J. Kim (ID)
Korea University
Seoul, Korea (Republic of)

Won-Ki Jeong (ID)
Korea University
Seoul, Korea (Republic of)

Angelica I. Aviles-Rivero (ID)
University of Cambridge
Cambridge, UK

Junjun He (ID)
Shanghai AI Laboratory
Shanghai, China

Shaoting Zhang (ID)
Shanghai AI Laboratory
Shanghai, China

ISSN 0302-9743   ISSN 1611-3349 (electronic)
Lecture Notes in Computer Science
ISBN 978-3-031-73470-0   ISBN 978-3-031-73471-7 (eBook)
https://doi.org/10.1007/978-3-031-73471-7

Preface

The Second International Workshop on Foundation Models for General Medical AI (MedAGI) was held in Marrakesh, Morocco on October 6, 2024, in conjunction with the 27th International Conference on Medical Image Computing and Computer-Assisted Intervention (MICCAI 2024).

Medical image analysis has traditionally relied on AI models trained on specific datasets, which often becomes challenging when transferred to data from different medical centers. This inherent limitation has inspired a growing interest in general medical AI, capable of seamlessly adapting to various medical scenarios, data modalities, and task formulations prevalent across hospitals and institutions. Drawing parallels from the computer vision and natural language processing domains, foundation models, such as large language and vision-language models like GPT and LLaMA, stand out as quintessential general AI solutions. These models have demonstrated remarkable proficiency in a myriad of tasks owing to their massive training datasets and substantial model sizes. Yet, the translation of these successes to medicine, namely general medical AI, remains nascent. This workshop was designed to continue the success of the 2023 event and serve as a confluence of insights from the current landscape of medical AI and foundation models.

In this year's workshop, we aimed to foster discussions that will pave the way for the evolution of task-specific medical AI systems into more generalized frameworks capable of tackling a diverse range of tasks, datasets, and domains.

The workshop received 32 submissions, including 26 full-length manuscripts and 6 extended abstracts. All the full-length manuscripts were peer-reviewed in a double-blind format. Each manuscript was reviewed by at least three reviewers who were experts in medical image analysis. The invaluable reviews from 40 reviewers contributed to the final decision: 8 papers were accepted for oral presentation and 9 for poster presentation. Both oral and poster papers were intended to be published in this Springer LNCS volume. The review process for extended abstracts was single-blind and was conducted by two workshop chairs. The extended abstracts are not published in this volume but will be archived on the workshop website.

The success of the workshop is due to the valuable work and significant efforts of all the authors, the reviewers, the invited speakers, and the attendees. We thank all of them for their contributions. We hope that this workshop will push forward the development of general medical AI for diverse tasks, datasets, and domains.

October 2024

Zhongying Deng
Yiqing Shen
Hyunwoo J. Kim
Won-Ki Jeong
Angelica I. Aviles-Rivero
Junjun He
Shaoting Zhang

Organization

Program Committee Chairs

Zhongying Deng   University of Cambridge, UK
Yiqing Shen   Johns Hopkins University, USA
Hyunwoo Kim   Korea University, South Korea
Won-Ki Jeong   Korea University, South Korea
Angelica I. Aviles-Rivero   University of Cambridge, UK
Junjun He   Shanghai AI Laboratory, China
Shaoting Zhang   Shanghai AI Laboratory, China

Program Committee

Amritpal Singh   Georgia Institute of Technology, USA
Chaoyan Huang   Chinese University of Hong Kong, China
Chong Wang   Beihang University, China
Dogyun Park   Korea University, South Korea
Eduard Lloret Carbonell   Shanghai Jiao Tong University, China
Erfan Darzi   UMCG, University of Groningen, Netherlands
Guannan He   Johns Hopkins University, USA
Hongtao Wu   Hong Kong University of Science and Technology (Guangzhou), China
Hyungyung Lee   KAIST GSAI, South Korea
Injae Kim   Korea University, South Korea
Jaejun Yoo   UNIST, South Korea
Jang-Hwan Choi   Ewha Womans University, South Korea
Jin Tae Kwak   Korea University, South Korea
Jinyoung Park   Korea University, South Korea
Jong Hak Moon   KAIST, South Korea
Jongha Kim   Korea University, South Korea
Joonmyung Choi   Korea University, South Korea
Lipei Zhang   University of Cambridge, UK
Sanghyeok Lee   Korea University, South Korea
Se Young Chun   Seoul National University, South Korea
Seong Jae Hwang   Yonsei University, South Korea
Seung Jun Baek   Korea University, South Korea
Seunghun Lee   Korea University, South Korea

Contents

FastSAM-3DSlicer: A 3D-Slicer Extension for 3D Volumetric Segment Anything Model with Uncertainty Quantification

Yiqing Shen[1], Xinyuan Shao[1], Blanca Inigo Romillo[1], David Dreizin[2], and Mathias Unberath[1(✉)]

[1] Johns Hopkins University, Baltimore, MD 21218, USA
{yshen92,unberath}@jhu.edu
[2] University of Maryland School of Medicine and R Adams Cowley Shock Trauma Center, Baltimore, MD 21201, USA

Abstract. Accurate segmentation of anatomical structures and pathological regions in medical images is crucial for diagnosis, treatment planning, and disease monitoring. While the Segment Anything Model (SAM) and its variants have demonstrated impressive interactive segmentation capabilities on image types not seen during training without the need for domain adaptation or retraining, their practical application in volumetric 3D medical imaging workflows has been hindered by the lack of a user-friendly interface. To address this challenge, we introduce FastSAM-3DSlicer, a 3D Slicer extension that integrates both 2D and 3D SAM models, including SAM-Med2D, MedSAM, SAM-Med3D, and FastSAM-3D. Building on the well-established open-source 3D Slicer platform, our extension enables efficient, real-time segmentation of 3D volumetric medical images, with seamless interaction and visualization. By automating the handling of raw image data, user prompts, and segmented masks, FastSAM-3DSlicer provides a streamlined, user-friendly interface that can be easily incorporated into medical image analysis workflows. Performance evaluations reveal that the FastSAM-3DSlicer extension running FastSAM-3D achieves low inference times of only 1.09 s per volume on CPU and 0.73 s per volume on GPU, making it well-suited for real-time interactive segmentation. Moreover, we introduce an uncertainty quantification scheme that leverages the rapid inference capabilities of FastSAM-3D for practical implementation, further enhancing its reliability and applicability in medical settings. FastSAM-3DSlicer offers an interactive platform and user interface for 2D and 3D interactive volumetric medical image segmentation, offering a powerful combination of efficiency, precision, and ease of use with SAMs. The source code and a video demonstration are publicly available at https://github.com/arcadelab/FastSAM3D_slicer.

Keywords: Foundation Model · Deep Learning · Segment Anything Model (SAM) · Interactive Segmentation · Interface · 3D Slicer

© The Author(s), under exclusive license to Springer Nature Switzerland AG 2025
Z. Deng et al. (Eds.): MedAGI 2024, LNCS 15184, pp. 1–9, 2025.
https://doi.org/10.1007/978-3-031-73471-7_1

1   Introduction

Precise segmentation of anatomical structures and pathological regions from medical images is essential for accurate diagnosis, treatment planning, and monitoring of disease progression [4]. However, manual segmentation is time-consuming, labor-intensive, and prone to inter-observer variability. Deep-learning-driven automatic segmentation models have shown promise in reducing manual effort, they often struggle to generalize across diverse datasets, anatomical variations, and unseen pathologies during inference [6]. The Segment Anything Model (SAM) [5] and its variants for volumetric medical images, such as SAM-Med3D [11], have emerged as flexible zero-shot solutions that enable interactive segmentation of novel objects without requiring retraining. These models are designed to generalize across various tasks and datasets through large-scale pre-training and support for manual prompting. The interactive nature of SAM-based segmentation makes fast inference times particularly important, as it allows users to make immediate adjustments and corrections, improving the accuracy and efficiency of the segmentation process. FastSAM-3D [10], a computationally efficient 3D SAM architecture, has been specifically optimized for real-time interactive segmentation of 3D volumes such as computed tomography (CT). FastSAM-3D utilizes a compact Vision Transformer (ViT) encoder [2], distilled from the larger SAM-Med3D [11], and incorporates an efficient 3D Sparse Flash Attention [10] mechanism to reduce computational costs while maintaining high segmentation quality.

Despite the potential of efficient SAM variants to enable highly responsive interactive segmentation, there is currently a lack of user-friendly interfaces to facilitate their practical application in medical image analysis workflow. Interactive prompting in 3D medical volumes is more challenging compared to 2D interfaces for natural images, due to the increased complexity of visualizing 3D data on a 2D screen and the need for seamless integration with existing medical image analysis workflows. 3D Slicer[1] [3] is an open-source software platform widely used for the analysis and visualization of volumetric medical images. It supports a variety of plug-ins and offers a robust framework for the integration of new tools, making it an ideal choice for creating an interface for interactive volumetric image segmentation, e. g., with FastSAM-3D. In this manuscript, we describe FastSAM-3DSlicer, a plugin for integrating 2D and 3D SAM models, including the efficient FastSAM-3D, into the well-established image analysis platform 3D Slicer. This extension enables users to load 3D volumes, interactively annotate structures of interest using point prompts, and visualize the resulting segmentations in real time within a familiar software environment.

In summary, our contributions are three-fold. Firstly, we propose a novel 3D Slicer-based extension for both 2D and 3D SAM models, including the efficient FastSAM-3D, for volumetric image segmentation. It enables interactive prompting in a 3D manner with SAM. Secondly, we show the quantitative comparison of the inference time of different SAM models within our interface on both

---

[1] https://www.slicer.org.

CPU and GPU environments. Finally, we propose an uncertainty quantification scheme based on the fast inference speed of FastSAM-3D in our extension. It can guide the user for better prompting.

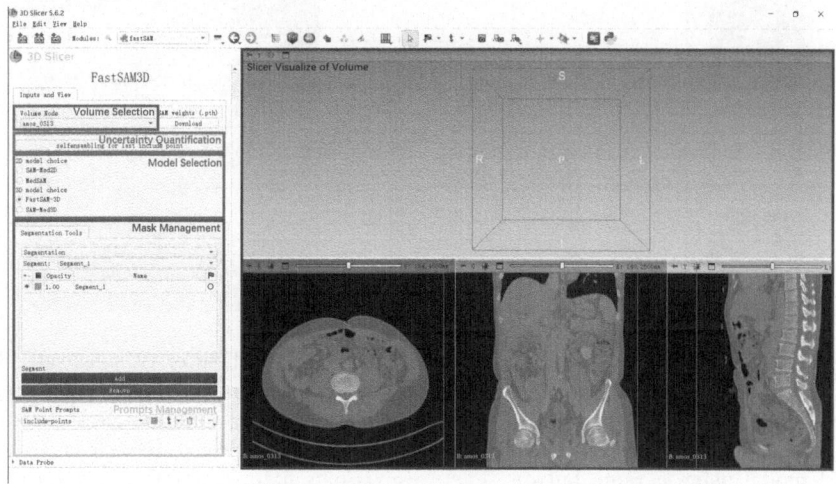

Fig. 1. An illustration of the interface of our 3D Slicer extension *i.e.*, FastSAM-3DSlicer. It allows users to load 3D volumes (red), select the mode for uncertainty quantification (green), choose SAM models (brown), manage masks (blue) and prompts (yellow), and visualize segmentation and input volume in real-time (purple). This integration enhances image analysis workflows by providing an intuitive and efficient interface for the interactive annotation and segmentation of volumetric medical images with SAM. (Color figure online)

2   Methods

Overview of the 3D Slicer Extension. The interface for our FastSAM-3DSlicer extension is illustrated in Fig. 1, with its overall implementation details shown in Fig. 2. Unlike previous works integrating 2D SAMs into 3D Slicer, such as TomoSAM [9] and SAMME [7], FastSAM-3D's efficiency at the volume level eliminates the need to prepare image embeddings before interactive segmentation (*i.e.*, user prompting) in 3D Slicer. This improvement enhances the user experience by reducing pre-processing time and allowing for more immediate interaction with the data. Upon importing a 3D volumetric image into FastSAM-3DSlicer, the extension automatically generates three node types, including a volume node, a segmentation node, and two markup nodes. These node types represent specific data structures within 3D Slicer that manage different aspects of the image and segmentation process. The volume node contains the raw 3D image in grayscale, represented as a NumPy array, which serves as the input for

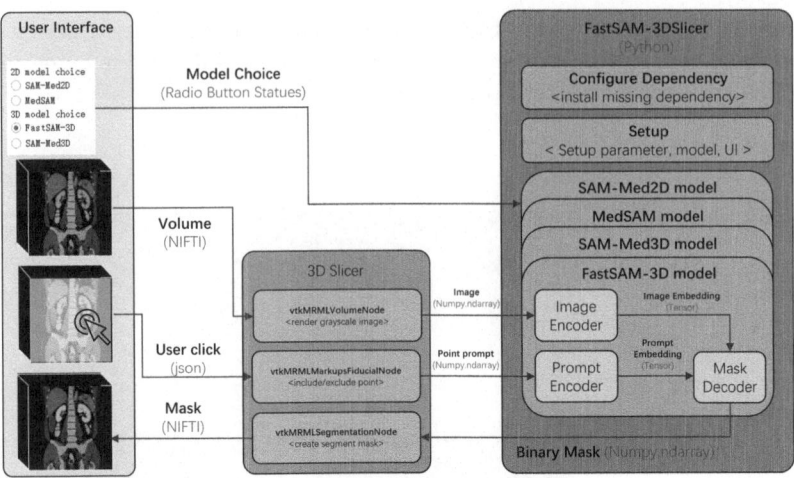

Fig. 2. The overall implementation workflow of FastSAM-3DSlicer. When FastSAM-3DSlicer is started for the first time, it runs the configure dependency block to check and install all required dependencies. The setup block runs every time FastSAM-3DSlicer is started to initialize parameters, models, and the user interface. Users can select a SAM model using a radio button. The green (Color figure online) blocks represent the components of the 3D Slicer environment, including the volume node, markup nodes, and segmentation node, which handle the raw image data, user interactions, and generated masks, respectively. The blue blocks represent the FastSAM-3DSlicer Python components, which include configuring dependencies, setup, and the various supported 2D and 3D SAM models (SAM-Med2D [1], MedSAM [8], SAM-Med3D [11], and FastSAM-3D [10]). Black text indicates processes or actions within the workflow, while gray text represents data types or storage formats used throughout the process.

the SAM model The segmentation node stores all the segmented masks, which are updated when users add new point prompts and generated when users perform the add mask operation. The two markup nodes hold all the include and exclude points that the user can add interactively. When a user adds an include or exclude point prompt, FastSAM-3DSlicer first converts this point from RAS to XYZ coordinates based on the affine matrix stored in the input NIfTI file. Using this new point prompt and previous input point prompts, FastSAM-3DSlicer crops or pads the raw image to match the selected SAM model's input size and feeds it into the image encoder to generate image embeddings in real time. The prompt encoder processes all the input points to generate prompt embeddings. The mask decoder then translates the image embeddings, prompt embeddings, and previous masks into a new mask. This segmentation mask is resized to the raw image dimensions based on saved coordinate information and updated in the segmentation node. In addition to supporting the FastSAM-3D model, FastSAM-3DSlicer also supports SAM-Med3D [11], MedSAM [8], and SAM-Med2D [1], all of which follow the same structural process.

Uncertainty Quantification Scheme. Uncertainty quantification in FastSAM-3DSlicer provides users with a measure of confidence in the segmentation results, which can be used to guide user prompting. Regions with higher uncertainty indicate a greater need for additional prompts. Our method leverages the efficiency of FastSAM-3D by running the image encoder once and performing multiple decoding steps to generate an ensemble of segmentations. It begins with the initial segmentation, where the image encoder processes the input 3D volumetric image $\mathbf{I}$ to produce its image embedding $\mathbf{E} = \text{Encoder}(\mathbf{I})$. Next, the mask decoder generates the initial segmentation logits $\widehat{\mathbf{M}}_0$ based on the image embedding $\mathbf{E}$ and the initial set of point prompts $(\mathbf{P}_0)$ provided from the user, *i.e.* $\widehat{\mathbf{M}}_0 = \text{Decoder}(\mathbf{E}, \mathbf{P}_0)$. The segmentation mask $\mathbf{M}_0$ is the binarized segmentation logits $\widehat{\mathbf{M}}_0$, obtained by applying a threshold τ:

$$\mathbf{M}_0 = \mathbb{1}(\widehat{\mathbf{M}}_0 > \tau), \tag{1}$$

where $\mathbb{1}(\cdot)$ is the indicator function. To quantify uncertainty, subsequent point prompts are sampled from the initial segmentation mask $\mathbf{M}_0$. These pseudo-point prompts are used to run the decoder multiple times, each time producing a slightly different segmentation mask due to variations in the sampled prompts. Let $\mathbf{P}_i$ denote the point prompts sampled from the initial segmentation mask $\mathbf{M}_0$ for the ith iteration: $\mathbf{P}_i = \text{SamplePrompts}(\mathbf{M}_0)$. The decoder is then run N times using these sampled prompts, while keeping the image encoder constant, to produce N different segmentation masks $\{\widehat{\mathbf{M}}_i\}_{i=1}^{N}$ with $\widehat{\mathbf{M}}_i = \text{Decoder}(\mathbf{E}, \mathbf{P}_i)$. Since the image encoder only runs once, the majority of the computational efficiency is preserved, as the decoder accounts for a smaller proportion of the total computation. Inspired by the self-ensembling [13], the segmentation logits from each decoder run are averaged to produce the final ensemble result $\widehat{\mathbf{M}}$:

$$\widehat{\mathbf{M}} = \frac{1}{N} \sum_{i=1}^{N} \widehat{\mathbf{M}}_i. \tag{2}$$

This averaging process not only provides a robust final segmentation mask but also allows for the calculation of uncertainty. Specifically, the variability among the N segmentation masks can be used to estimate the uncertainty, formally expressed as the standard deviation or variance of the logits at each voxel:

$$\text{Uncertainty}(x) = \sqrt{\frac{1}{N} \sum_{i=1}^{N} (\widehat{\mathbf{M}}_i(x) - \widehat{\mathbf{M}}(x))^2} \tag{3}$$

where x denotes a voxel in the 3D volume.

3   Experiments

Implementation Details. We implemented the proposed 3D Slicer extension using 3D Slicer version 5.6.2 and Python version 3.10. The experiments were conducted in two distinct environments to evaluate inference times. The first environment was a CPU-only setup utilizing a laptop-level AMD Ryzen 5 5500U CPU, while the second environment employed a GPU setup with one NVIDIA RTX 2060 GPU. Following previous work [10], we use the test set of *totalsegmentator* [12] to test the inference time. All data are prepared in NIfTI format for loading. Code for the 3D Slicer extension is available at https://github.com/arcadelab/FastSAM3D_slicer. A video demo for our 3D slicer extension is available at https://www.youtube.com/watch?v=oJ9ZhnPWqSs.

Table 1. Comparison of inference times for different SAMs on CPU and GPU in our 3D Slicer extension. Results in red indicate slice-level inference times, while results in blue indicate volume-level inference times.

Models	Inference Time w/ CPU	Inference Time w/ GPU
SAM-Med2D [1]	1.52 s per slice	0.52 s per slice
MedSAM [8]	48.9 s per slice	12.69 s per slice
SAM-Med3D [11]	7.75 s per volume	1.76 s per volume
FastSAM-3D [10]	1.09 s per volume	0.73 s per volume

Results for Inference Time. Table 1 presents a comparison of inference times for different SAMs with both CPU and GPU environments within our FastSAM-3DSlicer extension. The SAM-Med2D [1] exhibits an inference time of 1.52 s per slice on the CPU and 0.52 s per slice on the GPU. While SAM-Med2D [1] is effective for 2D slice-based segmentation, it fails to address volume-level segmentation in 3D Slicer, as the user needs to provide prompts for each slice individually, which results in longer inference time. MedSAM [8] shows considerably higher inference times, with 48.9 s per slice on the CPU and 12.69 s per slice on the GPU. The increased processing time is due to its higher resolution of 1024×1024 compared to the 256×256 resolution of SAM-Med2D [1]. Its high computational cost limits its practicality for real-time applications in image analysis settings. The SAM-Med3D [11], optimized for 3D segmentation, achieves 7.75 s per volume on the CPU and 1.76 s per volume on the GPU. This demonstrates a substantial improvement over MedSAM [8], particularly in GPU environments, making it a more viable option for volume-level segmentation. FastSAM-3D [10] significantly outperforms the other models in terms of inference speed ($p < 0.01$). It achieves 1.09 s per volume on the CPU and 0.73 s per volume on the GPU. This reduction in inference time highlights the efficiency and optimization of FastSAM-3D for real-time interactive segmentation of 3D volumes. FastSAM-3D demonstrates the lowest inference times on volume-level segmentation across both CPU and GPU environments, affirming its suitability for integration into image analysis workflows where speed and efficiency are critical. The improvements in inference time not only facilitate faster segmentation

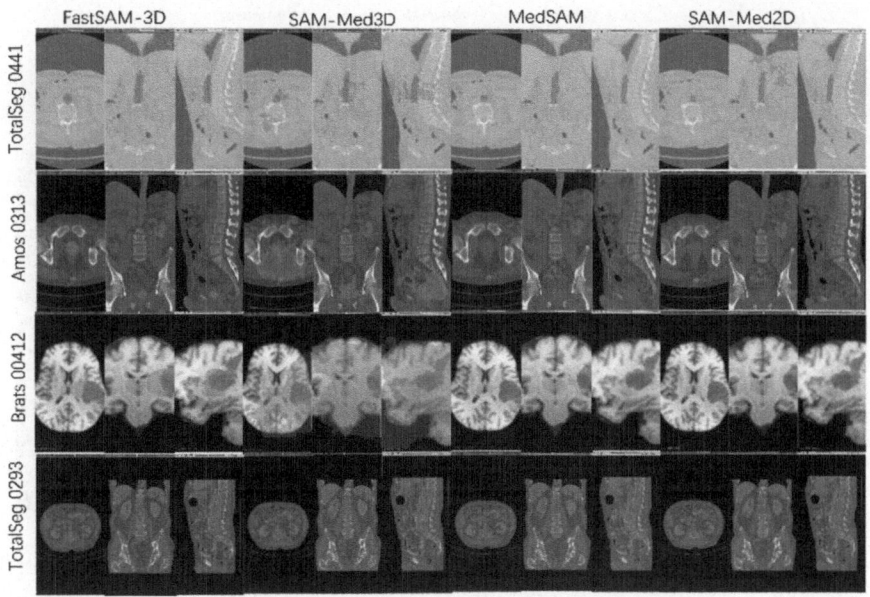

Fig. 3. Qualitative comparison of segmentation results from different SAM models in the FastSAM-3DSlicer interface. Each row showcases the segmentation performance on a specific anatomical structure or region of interest. The red (Color figure online) masks represent the ground truth segmentations, while the green masks depict the predicted segmentations from each model. (Color figure online)

but also enable more immediate and iterative interaction with the volumetric data, thereby enhancing the overall utility of the 3D Slicer extension in medical image analysis applications.

Illustrative Visual Examples. Figure 3 presents a qualitative comparison of segmentation results obtained from different SAM models integrated into the FastSAM-3DSlicer interface. The examples showcase the performance of each model on various anatomical structures and regions of interest. FastSAM-3D and SAM-Med3D generate segmentations for the entire 3D volume, demonstrating their ability to capture spatial context and produce coherent masks. In contrast, MedSAM and SAM-Med2D operate on 2D slices, resulting in smaller and more localized segmentations when visualized in the 3D view. Across all examples, FastSAM-3D exhibits the highest agreement with the ground truth, highlighting its superior performance in terms of both efficiency and accuracy for real-time 3D interactive segmentation within the FastSAM-3DSlicer.

Results for Uncertainty Quantification. Figure 4 showcases the uncertainty quantification results obtained through the self-ensembling approach in the FastSAM-3DSlicer extension. The results demonstrate that FastSAM-3D

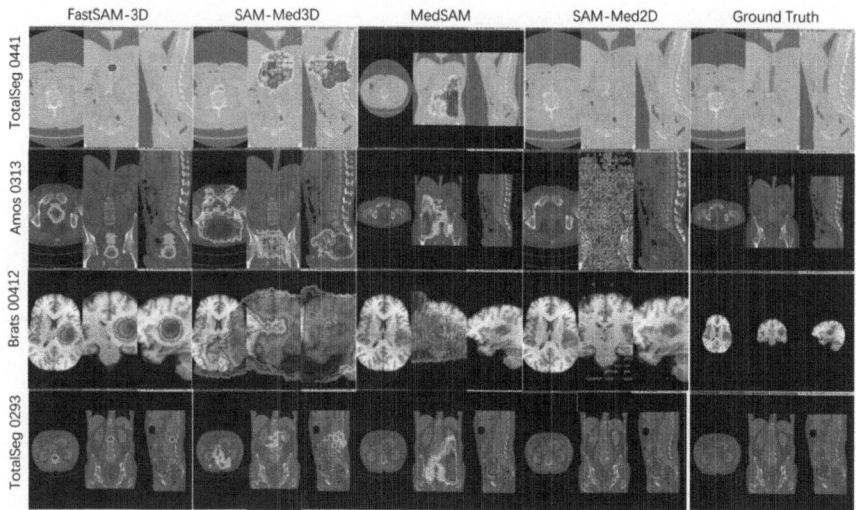

Fig. 4. Visualization of the uncertainty quantification results obtained through self-ensembling in the FastSAM-3DSlicer. Each column represents a different SAM model, while the last column shows the ground truth segmentation. The uncertainty maps are displayed as heatmaps overlaid on the original images, where lighter colors indicate higher uncertainty (lower probability) in the segmentation.

exhibits the lowest overall uncertainty among the compared models, highlighting its robustness and reliability for real-time 3D interactive segmentation. It allows for the estimation of uncertainty by calculating the variance among the ensemble of segmentations. The uncertainty information provided by FastSAM-3DSlicer is particularly valuable for guiding user interactions, as it identifies regions where additional user prompts may be required to improve segmentation accuracy. By focusing on areas of high uncertainty, users can iteratively refine the segmentation results, leading to more precise and reliable delineations of anatomical structures or regions of interest.

4   Conclusion

We presented FastSAM-3DSlicer, a 3D Slicer extension designed to facilitate real-time, interactive segmentation of 3D volumetric medical images with SAMs. Our extension integrates SAM-Med2D, MedSAM, SAM-Med3D, and FastSAM-3D, offering a user-friendly interface that automates the handling of user prompts and segmented masks within the familiar 3D Slicer environment. Our extension demonstrates superior performance in terms of inference time, particularly with the FastSAM-3D model, which achieves low inference times on both CPU and GPU environments. This makes FastSAM-3D highly suitable for real-time applications, reducing the computational burden while maintaining high segmentation quality. Furthermore, the integration of an innovative uncertainty quantification scheme leverages the rapid inference capabilities of FastSAM-3D, providing users with additional information about the reliability of the segmentation results. Overall, by combining computational efficiency, precision, and ease of use, FastSAM-3DSlicer addresses the need for a user-friendly interface for SAMs, thereby enhancing decision-making processes and improving patient outcomes in medical settings.

References

1. Cheng, J., et al.: Sam-Med2D. arXiv preprint arXiv:2308.16184 (2023)
2. Dosovitskiy, A., et al.: An image is worth 16×16 words: transformers for image recognition at scale. arXiv preprint arXiv:2010.11929 (2020)
3. Fedorov, A., et al.: 3D slicer as an image computing platform for the quantitative imaging network. Magn. Reson. Imaging **30**(9), 1323–1341 (2012)
4. Isensee, F., et al.: nnU-Net: a self-configuring method for deep learning-based biomedical image segmentation. Nat. Methods **18**(2), 203–211 (2021)
5. Kirillov, A., et al.: Segment anything. arXiv preprint arXiv:2304.02643 (2023)
6. Liu, X., et al.: Review of deep learning based automatic segmentation for lung cancer radiotherapy. Front. Oncol. **11**, 717039 (2021)
7. Liu, Y., et al.: Segment any medical model extended. In: Medical Imaging 2024: Image Processing, vol. 12926, pp. 411–422. SPIE (2024)
8. Ma, J., et al.: Segment anything in medical images. Nat. Commun. **15**(1), 654 (2024)
9. Semeraro, F., et al.: TomoSAM: a 3D slicer extension using SAM for tomography segmentation. arXiv preprint arXiv:2306.08609 (2023)
10. Shen, Y., et al.: Fastsam3D: an efficient segment anything model for 3D volumetric medical images. arXiv preprint arXiv:2403.09827 (2024)
11. Wang, H., et al.: Sam-Med3D. arXiv preprint arXiv:2310.15161 (2023)
12. Wasserthal, J., et al.: TotalSegmentator: robust segmentation of 104 anatomic structures in CT images. Radiol. Artif. Intell. **5**(5) e230235 (2023)
13. Zhang, Y., Hu, S., Jiang, C., Cheng, Y., Qi, Y.: Segment anything model with uncertainty rectification for auto-prompting medical image segmentation. arXiv preprint arXiv:2311.10529 (2023)

The Importance of Downstream Networks in Digital Pathology Foundation Models

Gustav Bredell, Marcel Fischer, Przemyslaw Szostak,
Samaneh Abbasi-Sureshjani⬤, and Alvaro Gomariz(✉)⬤

F. Hoffmann-La Roche AG,Basel, Switzerland
alvaro.gomariz@roche.com

Abstract. Digital pathology has significantly advanced disease detection and pathologist efficiency through the analysis of gigapixel whole-slide images (WSI). In this process, WSIs are first divided into patches, for which a *feature extractor* model is applied to obtain feature vectors, which are subsequently processed by an *aggregation model* to predict the respective WSI label. With the rapid evolution of representation learning, numerous new feature extractor models, often termed foundational models, have emerged. Traditional evaluation methods rely on a static downstream aggregation model setup, encompassing a fixed architecture and hyperparameters, a practice we identify as potentially biasing the results. Our study uncovers a sensitivity of feature extractor models towards aggregation model configurations, indicating that performance comparability can be skewed based on the chosen configurations. By accounting for this sensitivity, we find that the performance of many current feature extractor models is notably similar. We support this insight by evaluating seven feature extractor models across three different datasets with 162 different aggregation model configurations. This comprehensive approach provides a more nuanced understanding of the feature extractors' sensitivity to various aggregation model configurations, leading to a fairer and more accurate assessment of new foundation models in digital pathology.

1 Introduction

Digital pathology (DP) has significantly advanced with automated solutions for tasks like breast cancer [7] and metastases detection [15], leveraging gigapixel whole-slide images (WSI) stained with H&E. The challenge of applying standard deep learning models for processing these large images has led to the adoption of the multiple instance learning (MIL) framework. In MIL, as depicted in step 1 in Fig. 2, WSIs are divided into patches, also known as tiles. A *feature extractor* model extracts features from each tile to generate embedding vectors. These vectors, collectively referred to as a *bag*, are then processed by an *aggregation*

Supplementary Information The online version contains supplementary material available at https://doi.org/10.1007/978-3-031-73471-7_2.

model to predict the WSI label [9,16]. Popular choices for aggregation models include AttentionMIL [13] and TransMIL [18], which both rely on using attention mechanisms for feature aggregation.

Beyond computational efficiency, feature extractors play a critical role in overcoming the scarcity of labeled data in DP. Using representation learning approaches feature extractors can be trained on large datasets of unlabeled images enabling their use across diverse datasets. Since the pivotal work of Chen et al. [5], which significantly improved visual representations using contrastive learning (SimCLR), a range of novel representation learning approaches has been introduced. SimCLR learns representations by ensuring that the embeddings of images with the same label (positive examples) are close, whereas the embeddings of images with different labels (negative examples) are far apart. Subsquently, Grill et al. [11] showed that self-supervised learning can also be done without negative examples (BYOL). This approach was further improved and combined with transformers leading to DINO [3]. The most recent approaches combine masked autoencoder (MAE) [12] with self-distillation. This is the strategy used by iBOT [23] and is also at the core of DINOv2 [17].

The DP field has adapted these representation learning advancements, notably in CTransPath [21] and REMEDIS [1]. Due to the large number of tiles that can be extracted from a single WSI and the availability of large publicly available datasets, such as TCGA [19], the datasets for CTransPath and REMEDIS contain 16 Mio and 50 Mio tiles, respectively. These large datasets allow the development of superior feature extractors, now commonly known as *foundational models*, that generalize across datasets without the need for retraining. Filiot et al. [8] made a first step in this direction and demonstrated better classification performance compared to CTransPath by extracting feature embeddings using a model trained with iBOT. Chen et al. [4] went a step further and increased the dataset size to 100 Mio tiles while using DINOv2 to train the feature extraction model. Finally, one of the most recent feature extractors, Virchow [20], was also trained using the DINOv2 approach but on a dataset size of 380 Mio tiles.

Foundational models promise to extract informative features from patches across diverse datasets. Ideally, capturing relevant features enhances downstream tasks, such as classification, while poor features hinder it. Feature extractors are often evaluated through their performance in basic classification tasks using models such as linear or K-NN classification [6]. However, in digital pathology, the use of an aggregation model to process embedding vectors and make final predictions can complicate the assessment of the feature extraction quality.

As illustrated with an example in Fig. 1, our analysis reveals that, whereas foundational models do indeed have some influence on the classification performance, they are highly sensitive to the aggregation model configuration. Thus, when comparing feature extractors, the sensitivity to the second step of the classification pipeline, namely the aggregation model, is an important variable to control for. Our contribution in this paper is twofold.

- We characterize the feature extractors' sensitivity to various aggregation model configurations, challenging traditional feature extractor evaluation methods in digital pathology.
- We propose a framework for stringent and fair evaluation for state-of-the-art feature extractors.

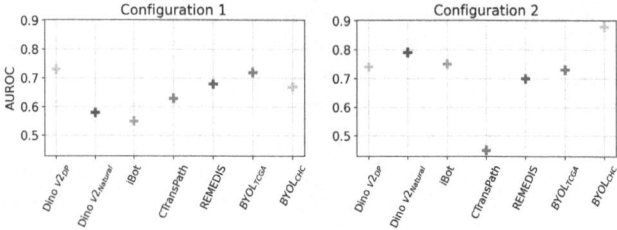

Fig. 1. Typical frameworks for evaluation of feature extraction models use fixed configurations in the aggregation models, leading to substantially different results and hence limited informative value.

2   Experimental Setup and Methods

This section outlines the classification pipeline for whole-slide images (WSIs) and outlines the framework we use for evaluating the sensitivity of feature extractors towards aggregation model configurations.

2.1   Pipeline for Classification of WSIs

As depicted in Fig. 2, the typical MIL pipeline for DP requires two models to obtain a final classification for a given WSI. First, a feature extraction model leverages recent self-supervised learning advancements and extensive datasets to generalize across tasks and datasets [2,4,20]. This model is applied to tiles in a WSI to produce feature embedding vectors. Next, a smaller aggregation model, specific to each dataset, processes the extracted embeddings to aggregate information and classify the WSI. In contrast to the feature extraction model, this aggregation model is re-trained for each dataset.

Feature extraction models have been compared under a single aggregation model configuration, i.e. fixed model architecture and hyperparameters [1,4,14]. Figure 1 illustrates the significant impact of aggregation model configuration choice on performance, rendering widely adopted evaluation frameworks suboptimal. Indeed, fixed aggregation model configurations can favour some feature extraction models while penalizing others. We outline an experimental setup to thoroughly address two critical questions:

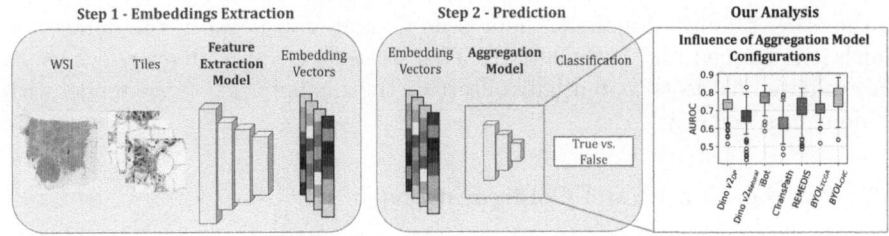

Fig. 2. Illustration of the typical classification pipeline with MIL in digital pathology.

Question 1: Can a single aggregation model configuration optimally support various feature extraction models?

Question 2: How do state-of-the-art feature extraction models perform relative to each other when controlling for different aggregation model configurations?

2.2 Feature Extraction Models

To explore our research questions, we assess seven feature extraction models, with details and characteristics provided in Supplementary Table S1.

We begin by evaluating the ViT-L model from DINOv2, trained on 142 million natural images with 300 million parameters, to assess the applicability of models trained on natural images for DP [17]. Additionally, we explore recently published models specifically designed for DP: CTransPath [21], REMEDIS [1], and iBOT [8] (teacher model), all trained on extensive DP datasets as detailed in Supplementary Table S1. Lastly, we investigate three feature extracting models trained in-house.

We train from scratch a DINOv2 ViT-L model on TCGA and a vast in-house dataset with diverse tissue types and real-world data. The total training dataset size is 35 million 224×244 tiles. WSIs are usually acquired at different magnifications. Tiles from $20\times$ magnification offer broader content, while $40\times$ magnification tiles provide finer tissue details due to their higher resolution. We train at both $20\times$ and $40\times$ magnifications to capture different features, following successful strategies in the literature [14]. We employ the official DINOv2 repository with the the default parameters for ViT-L/16 training, with a few exceptions. We decrease the batch size to 352 due to computational constraints. Due to the decreased batch size, we increase the number of epochs to 270, warm-up epochs to 25 and adjust the learning rate to 1.375×10^{-3} according to the heuristic by Goyal et al. [10].

Two ResNet-50 models are trained using BYOL: The first, $BYOL_{TCGA}$, utilizes 2 million tiles randomly sampled from the TCGA dataset at $20\times$ magnification, featuring a smaller dataset and model size for comparison. The second, $BYOL_{CHC}$ (according to the first letter of each of the three evaluation datasets), is also trained on 2 million tiles but randomly sampled from the training set of the evaluation datasets. Thus even though the training dataset is small compared to the other published models, there is no domain gap between the dataset on which it is trained and evaluated on. This strategy ensures direct relevance to the evaluated datasets, potentially offsetting the smaller scale of the model with its domain specificity.

2.3   Aggregation Model Configurations

To investigate the performance fluctuation of the MIL pipeline when the feature extraction model is fixed and the aggregation model configuration change, we use different network hyperparameters and two well adopted aggregation model architectures: AttentionMIL [13], which uses an attention mechanism to aggregate tile information and assumes no interdependency between the tiles. TransMIL [18], which learns inter-tile dependencies by using the self-attention mechanism of transformers, in particular the Nyströmformer [22]. We change four hyperparameters with three distinct values each as shown in Table 1. These are decided heuristically with preliminary experiments assessing the influence and effective range of each hyperparameter. The resulting 162 different configurations (81 for each of the 2 architectures) are outlined below.

Table 1. Set of hyperparameter values for each aggregation model. Layers refer to fully connected layers in AttentionMIL and to attention blocks in TransMIL.

Hyperpar.	AttentionMIL	TransMIL
Learn. rate	$1e-4$, $1e-3$, $1e-2$	$1e-5$, $1e-4$, $1e-3$
Bag size	128, 1024, 8192	128, 1024, 2048
Layers	(512), (512, 384, 384), (512, 256, 128, 64, 32)	1, 2, 3
Dropout	0.00, 0.25, 0.50	0.00, 0.25, 0.50

AttentionMIL: When creating a batch for the aggregation model during training, there are two relevant parameters. One is the *bag size*, which determines the amount of tiles that is sampled from a particular WSI. The second is the *bags per batch*, determining how many bags from different WSIs are collected to form a batch. Here, we vary only the bag size parameter since it showed a larger influence. The final batch size=*bag size*bags per batch*. The *Layers* parameter corresponds to the number of nodes in the fully connected (FC) layers in the aggregation model. The list of numbers in Table 1 indicate the number of nodes for each layer. Lastly, the dropout parameter refers to the dropout which is applied at every layer of the aggregation model.

TransMIL: The selected hyperparameters for TransMIL are different due to the model architecture being a transformer, which does not employ FC layers. *Layers* refers to the number of Nyströmformer attention blocks. We also reduce the maximal bag size to 2048 due to computational limitations.

Both models share fixed training parameters: a weight decay of 10^{-5}, four *bags per batch*, AdamW optimizer, weighted cross entropy loss, and a cosine annealing scheduler. Aggregation models are trained for 50 epochs to ensure convergence within our configuration range.

2.4   Evaluation Datasets

Our study evaluates binary classification performance of feature extraction models across three distinct DP datasets. Thereby providing a more generalizable answer to our research questions. These datasets, comprising H&E-stained histopathology slides WSIs, allow us to assess each feature extractor under 162 different aggregation model configurations. This comprehensive approach, covering 7 feature extractors, 162 aggregation model configurations, and 3 datasets, culminates in a total comparison of $7 \times 162 \times 3 = 3402$ experimental configurations.

Performance metrics include the area under the receiver operating characteristic curve (AUROC) and average precision (AP), both ranging from 0 to 1. A higher AUROC indicates superior distinction between the positive and negative classes, while a higher AP reflects more accurate predictions of positive instances across all recall levels, effectively balancing precision and recall. Performance metrics are derived from the test set, using the aggregation model's epoch with best validation score during training.

COO: Binary classification of cell of origin (COO). Each image contains the COO prediction label of activated B-cell like (ABC) or germinal center B-cell like (GCB) tumors in diffuse large B-cell lymphoma (DLBCL). 709 WSIs from two internal datasets were used. This data closely mirrors real-world data, since it is crucial to assess classification approaches in DP using such data and tasks. The WSIs ($40\times$ magnification) have been scanned by Ventana DP200 scanners. The artifact-free tissue tiles of this dataset were combined and randomly split into 70% training set, 15% validation set and 15% test set.

Camelyon16: Binary classification of cancer metastases vs. healthy in H&E images of lymph node tissue. The Camelyon16 dataset [15] consists of 400 WSIs of sentinel lymph nodes. The dataset is publicly available. For our evaluation, all artifact-free tissue tiles were used as well as the official train-test split. 20% of the training data was used as the validation set.

Herohe: Binary classification of breast cancer human epidermal growth factor receptor 2 (HER2) using the publicly available Herohe [7] dataset. Each H&E stained WSI is either labeled as HER2 positive or HER2 negative. We use the artifact-free tiles from tumor regions detected with an in-house tumor segmentation model. The 508 WSIs are split according to the official train-test split. 20% of the training data is used as the validation set.

3   Results

This section addresses our initial inquiries, first assessing if a universally optimal aggregation model configuration exists for multiple feature extraction (foundation) models, and then comparing different feature extractors considering performance variability across aggregation model setups.

3.1   Aggregation Model Configuration Influence

Our analysis begins by evaluating the sensitivity of feature extractors, or foundation models, to various aggregation model configurations.

The heatmap in Fig. 3 displays the classification performance across all aggregation model configurations. Trends are consistent across both AUROC and AP scores. The heatmap legend aids in identifying patterns, such as configurations with the lowest learning rate positioned on the left of each feature aggregator, which tend to yield lower performance in the COO dataset when using CTransPath but not for other features extractors. Analysis of these heatmaps reveals:

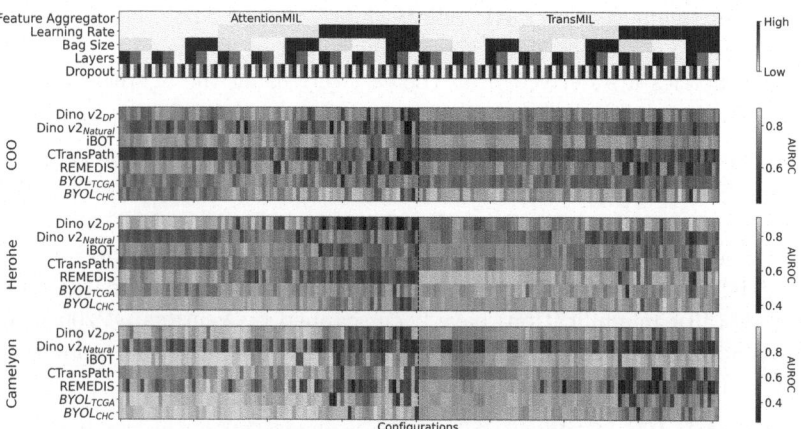

Fig. 3. The heatmap shows the performance of every aggregation model configuration set for each feature extraction model. The red colored legend shows how the configurations are ordered on the heatmap. (Color figure online)

Lack of a Universal Configuration: No single aggregation model configuration consistently outperforms across all feature extractors, as indicated by the absence of a uniformly bright column across models.

Dataset-Specific Configurations: Optimal configurations for a given feature extractor vary by dataset. While certain parameters like learning rate for AttentionMIL and the number of attention blocks for TransMIL show some dataset-specific importance, no definitive pattern emerges across datasets or models, suggesting the need for investigation of model-specific configurations.

These results highlight the need for evaluating a diverse range of configurations in the aggregation model. This approach would ascertain that any observed superiority of one feature extraction model over another is not simply attributed to the specific aggregation model setup selected.

3.2   Feature Extractor Comparison

Figure 4 diverges from the standard practice of showing a single outcome for a fixed aggregation model setup by presenting feature extractor model performance across all 162 configurations for various datasets. Through box plots, we observe substantial performance overlap among feature extraction models despite the variance across configurations. Key insights include:

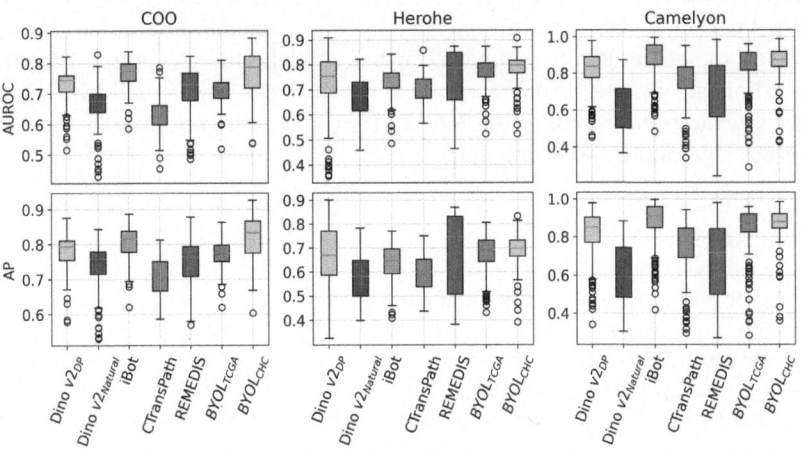

Fig. 4. Comparison of 7 feature extraction models across 162 different aggregation model configurations, which include 2 architectures with 81 parameters each.

Training on DP Datasets is Necessary: The DINOv2 model trained on natural images performs poorly compared to all other models, consistently for all the datasets.

Comparable Performance Across Model Sizes: The relatively small model $BYOL_{TCGA}$ matches the performance of larger ones, suggesting that larger models are not necessarily better for DP. This echoes Filiot et al.'s [8] findings that a ViT-B model can outperform a ViT-L model.

Feature Extractors Generalize Well: The $BYOL_{CHC}$ model, trained on WSIs from evaluation datasets, shows good performance across all datasets as expected. Interestingly, its performance is not much higher than that of other models such as $DINOv2_{DP}$, iBOT and $BYOL_{TCGA}$. This observation confirms that the feature extraction models have the capability to generalize well.

4 Conclusion

In this study, we challenge the prevailing methodology for comparing foundation models in digital pathology literature, demonstrating that it may yield misleading results. We show that due to the high sensitivity of feature extraction models to downstream aggregation model configurations, relying solely on a single aggregation model configuration can disproportionately favor certain feature extractor models while disadvantaging others. Hence, we propose evaluating foundation models across different configurations for fairer comparisons. Our comprehensive analysis, taking into account performance variations across multiple configurations of the aggregation model, reveals a considerable overlap in performance between different foundation or feature extractor models. Significantly, we find no universal aggregation model configuration that is uniformly effective for all feature extractors. Our work is limited though by only looking at classification tasks. In addition, the DINOv2 model we trained on digital pathology images might be subpar to other models due to computational and dataset limitations. Nevertheless, we believe this work will contribute to a more nuanced evaluation of foundation models that will help gain insight and further accelerate this rapidly evolving field.

References

1. Azizi, S., et al.: Robust and data-efficient generalization of self-supervised machine learning for diagnostic imaging. Nat. Biomed. Eng. **7**, 756–779 (2023)
2. Bommasani, R., et al.: On the opportunities and risks of foundation models. arXiv preprint arXiv:2108.07258 (2021)
3. Caron, M., et al.: Emerging properties in self-supervised vision transformers. In: Proceedings of the IEEE/CVF International Conference on Computer Vision, pp. 9650–9660 (2021)
4. Chen, R.J., et al.: A general-purpose self-supervised model for computational pathology. arXiv preprint arXiv:2308.15474 (2023)
5. Chen, T., Kornblith, S., Norouzi, M., Hinton, G.: A simple framework for contrastive learning of visual representations. In: International Conference on Machine Learning, pp. 1597–1607. PMLR (2020)
6. Chen, T., Kornblith, S., Swersky, K., Norouzi, M., Hinton, G.E.: Big self-supervised models are strong semi-supervised learners. Adv. Neural. Inf. Process. Syst. **33**, 22243–22255 (2020)
7. Conde-Sousa, E., et al.: Herohe challenge: predicting her2 status in breast cancer from hematoxylin&eosin whole-slide imaging. J. Imaging **8**(8) (2022). https://doi.org/10.3390/jimaging8080213, https://www.mdpi.com/2313-433X/8/8/213
8. Filiot, A., et al.: Scaling self-supervised learning for histopathology with masked image modeling. medRxiv, pp. 2023–07 (2023)
9. Gadermayr, M., Tschuchnig, M.: Multiple instance learning for digital pathology: a review on the state-of-the-art, limitations & future potential. arXiv preprint arXiv:2206.04425 (2022)
10. Goyal, P., et al.: Accurate, large minibatch SGD: training imagenet in 1 hour. arXiv preprint arXiv:1706.02677 (2017)

11. Grill, J.B., et al.: Bootstrap your own latent-a new approach to self-supervised learning. Adv. Neural. Inf. Process. Syst. **33**, 21271–21284 (2020)
12. He, K., Chen, X., Xie, S., Li, Y., Dollár, P., Girshick, R.: Masked autoencoders are scalable vision learners. In: Proceedings of the IEEE/CVF Conference on Computer Vision and Pattern Recognition, pp. 16000–16009 (2022)
13. Ilse, M., Tomczak, J., Welling, M.: Attention-based deep multiple instance learning. In: International Conference on Machine Learning, pp. 2127–2136. PMLR (2018)
14. Kang, M., Song, H., Park, S., Yoo, D., Pereira, S.: Benchmarking self-supervised learning on diverse pathology datasets. In: Proceedings of the IEEE/CVF Conference on Computer Vision and Pattern Recognition, pp. 3344–3354 (2023)
15. Litjens, G., et al.: 1399 H&E-stained sentinel lymph node sections of breast cancer patients: the CAMELYON dataset. GigaScience **7**(6), giy065 (2018). https://doi.org/10.1093/gigascience/giy065
16. Maron, O., Lozano-Pérez, T.: A framework for multiple-instance learning. In: Advances in Neural Information Processing Systems, vol. 10 (1997)
17. Oquab, M., et al.: DINOv2: learning robust visual features without supervision. arXiv preprint arXiv:2304.07193 (2023)
18. Shao, Z., Bian, H., Chen, Y., Wang, Y., Zhang, J., Ji, X., et al.: TransMIL: transformer based correlated multiple instance learning for whole slide image classification. Adv. Neural. Inf. Process. Syst. **34**, 2136–2147 (2021)
19. Tomczak, K., Czerwińska, P., Wiznerowicz, M.: Review The Cancer Genome Atlas (TCGA): an immeasurable source of knowledge. Contemp. Oncology/Współczesna Onkologia **2015**(1), 68–77 (2015)
20. Vorontsov, E., et al.: Virchow: a million-slide digital pathology foundation model. arXiv preprint arXiv:2309.07778 (2023)
21. Wang, X., et al.: Transformer-based unsupervised contrastive learning for histopathological image classification. Med. Image Anal. **81**, 102559 (2022)
22. Xiong, Y., et al.: Nyströmformer: a nyström-based algorithm for approximating self-attention. In: Proceedings of the AAAI Conference on Artificial Intelligence, vol. 35, pp. 14138–14148 (2021)
23. Zhou, J., et al.: iBoT: image BERT pre-training with online tokenizer. arXiv preprint arXiv:2111.07832 (2021)

Temporal-Spatial Adaptation of Promptable SAM Enhance Accuracy and Generalizability of Cine CMR Segmentation

Zhennong Chen, Sekeun Kim, Hui Ren, Quanzheng Li, and Xiang Li(✉)

Center of Advanced Medical Computing and Analysis, Massachusetts General Hospital and Harvard Medical School, Boston, MA 02114, USA
xli60@mgh.harvard.edu

Abstract. Accurate myocardium segmentation across all phases in one cardiac cycle in cine cardiac magnetic resonance (CMR) scans is crucial for comprehensively cardiac function analysis. Despite advancements in deep learning (DL) for automatic cine CMR segmentation, generalizability on unseen data remains a significant challenge. Recently, the segment-anything-model (SAM) has been invented as a segmentation foundation model, known for its accurate segmentation and more importantly, zero-shot generalization. SAM was trained on two-dimensional (2D) natural images; to adapt it for comprehensive cine CMR segmentation, we propose cineCMR-SAM which incorporates both temporal and spatial information through a modified model architecture. Compared to other state-of-the-art (SOTA) methods, our model achieved superior data-specific model segmentation accuracy on the STACOM2011 when fine-tuned on this dataset and demonstrated superior zero-shot generalization on two other large public datasets (ACDC and M&Ms) unseen during fine-tuning. Additionally, we introduced a text prompt feature in cineCMR-SAM to specify the view type of input slices (short-axis or long-axis), enhancing performance across all view types. The GitHub repository is https://github.com/zhennongchen/cineCMR-SAM.git.

Keywords: Cardiac magnetic resonance · Segmentation · Foundational Model

1 Introduction

Accurate and reproducible assessment of the myocardium is crucial to indicate previous infarcts, cardiomyopathies, or inflammatory diseases [1]. Cine cardiac magnetic resonance (CMR) imaging is considered as the gold standard modality for myocardial anatomy and function, but to quantify these metrics requires manual segmentation, a labor-intensive task hampered by variations in image quality and observer education and experience [2]. Moreover, conditions like cardiac dyssynchrony and diastolic heart failure demand segmentation across all cardiac phases for a more thorough temporal analysis, significantly extending segmentation time. Thus, there is a critical need for an automated cine CMR segmentation method capable of accurately segmenting all cardiac cycle phases in clinical practice.

Z. Chen and S. Kim—Co-first authors

© The Author(s), under exclusive license to Springer Nature Switzerland AG 2025
Z. Deng et al. (Eds.): MedAGI 2024, LNCS 15184, pp. 20–29, 2025.
https://doi.org/10.1007/978-3-031-73471-7_3

There have been advancements in DL for automatic cine CMR segmentation. Specifically, cine CMR segmentation here refers to "2D+T" CMR segmentation, which involves segmenting myocardium in one slice across all cardiac phases *simultaneously*. This requires DL models to understand both spatial information and temporal dynamics and maintain temporal consistency across all cardiac phases. Current mainstream approaches incorporate temporal information in cine CMR using three ways: (1) adding recurrent layers in a 2D convolutional neural network (CNN) to encode sequential information along the temporal axis [3–5], (2) utilizing a 3D CNN with 3D kernels that treat the temporal axis as a depth dimension [6], and (3) employing attention mechanisms on both temporal and spatial axes [7–9]. There are also studies [9, 10] that have addressed volumetric 3D+T segmentation, but we focus on 2D+T segmentation since segmenting in 2D slices allows networks to work with images even if they have different slice thicknesses or severe inter-slice misalignment due to cardiac or respiratory motion [11]. Despite these advancements, a major challenge remains in generalizability to unseen datasets. Several research [12–14] have demonstrated significant performance degradation when models are applied to new, unseen datasets from different centers or vendors. Another limitation is most DL methods are only validated on short-axis (SAX) views [15], but segmentation in long-axis (LAX) views is critical as it provides clinically valuable parameters such as longitudinal strains.

Recently, Segment-Anything-Model (SAM) [16] has been introduced as a segmentation foundation model trained on one-billion-image dataset. It is renowned for its segmentation accuracy and more importantly, zero-shot generalization and user-defined prompt input. Fine-tuning SAM on general medical images has outperformed specialist CNN [17, 18], and adapting SAM to specific image modalities has further enhanced accuracy considerably [19, 20]. However, there is no study tailoring SAM for cine CMR segmentation to leverage its generalization ability and prompt inputs. Therefore, we propose cineCMR-SAM with specific model modifications for cine CMR.

Our contributions are three-folds. First, we integrate the 2D SAM model for 2D+T CMR segmentation. Specially, we incorporate time-space self-attention in the SAM vision transformer (ViT) encoders, enabling extraction of both temporal and spatial information. Second, we design a text prompt feature to specify the view type of input slices (SAX or LAX) to enhance the segmentation across all views. Third, we fine-tune our model on one multi-center, multi-vendor public dataset (STACOM2011) and then demonstrate its superior zero-shot generalization compared to state-of-the-art (SOTA) methods on two other large datasets (ACDC and M&Ms) unseen during fine-tuning.

2   Methods

Figure 1 illustrates our model. In this section, we first present how we integrate a 2D SAM model for 2D+T segmentation. Then we introduce the prompt feature to specify the view type. Lastly, we cover some other model modifications and training procedures.

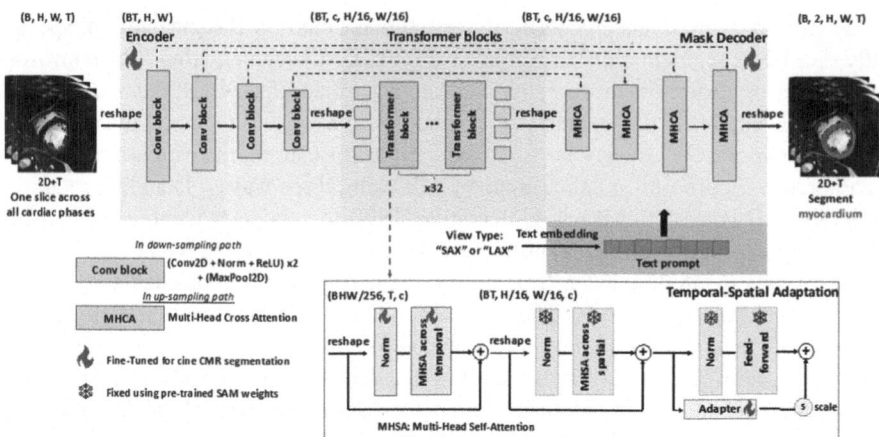

Fig. 1. The overall model structure. It includes (1) refined SAM ViT blocks with temporal-spatial adaptation (blue section); (2) U-Net framework as multi-scale encoder and decoder (green section); (3) text prompts to specify the view type (pink section). Components undergoing fine-tuning for CMR segmentation are denoted by a "fire" icon, whereas those retaining pre-trained SAM weights are marked with a "snowflake" icon. The data dimension before input into each module is annotated at the top of the respective module, where B = batch size, c = feature channel number, H and W = height and width of the input slice, and T = the number of cardiac phases. (Color figure online)

2.1 Integrate 2D SAM for 2D+T Segmentation

The SAM initially has three main parts: an image encoder, a prompt encoder and a mask decoder, all pre-trained on a vast one-billion-image dataset. The image encoder features a series of ViT blocks, critical for SAM's effectiveness in downstream tasks. Our cineCMR-SAM uses the "vit-h" variant with 32 transformer blocks. The initial design of SAM ViT blocks is for 2D images, extracting only spatial information using a multi-head self-attention (MHSA) mechanism. To adapt it for 2D+T segmentation, we were inspired by TimeSformer (the space-time self-attention module) [7] and thus refine the ViT blocks to have temporal MHSA and spatial MHSA sequentially applied one after the other. Skip connections are used to integrate temporal-spatial information. The layers in one refined ViT block are shown in the blue box in Fig. 1.

Regarding data dimension change through the model framework, for model inputs, we extract a CMR slice spanning across one cardiac cycle $x = \{x_t\}_{t=1}^T, x \in \mathbb{R}^{B \times H \times W \times T}$. Here, B denotes the batch size, $H \times W$ denote the slice dimensions and T denotes the number of phases in one cardiac cycle. Before the inputs are passed into the SAM backbone, a reshape operation is applied to transform $x \in \mathbb{R}^{B \times H \times W \times T}$ into $x \in \mathbb{R}^{BT \times H \times W}$ by merging the phases into the batch dimension. Then for ViT feature extraction, prior to feeding into the temporal MHSA, they are reshaped from $[BT, H/16, W/16, c]$ to $[BWH/256, T, c]$ so that the attention works on the temporal axis. Here c denotes feature channel number, while $H/16$ and $W/16$ denote the spatial dimensions of feature maps, which are down-sampled by 16 times because of the patch embedding process in transformer. After temporal MHSA, we transform the dimension from $[BWH/256, T, c]$ back to $[BT, H/16, W/16, c]$ for spatial MHSA. Using these dimension transformation operations, we can apply attention on both the temporal and spatial axes.

2.2 Text Prompt to Specify View Type

SAM allows users to input prompts to guide segmentation. Our goal is to enable cineCMR-SAM to segment both SAX and LAX using a *single* model. We hypothesize that leveraging the text prompt feature to specify the view type of the model input can enhance segmentation accuracy across all views. Concretely, we use text "SAX" and "LAX" as text prompts for SAX and LAX view inputs respectively. The text is embedded using the trainable prompt encoder and integrated into the decoder part of the model. Currently the prompt context is defined manually for each input, but it can be automated using deep learning [21].

2.3 Other Modifications and Training Procedure

As shown in Fig. 1, we utilize a 2D U-Net framework for our model and incorporate the refined SAM's ViT blocks at the bottom of U-Net. We hypothesize that using U-Net can better extract multi-scale spatial features in the image while the transformer can learn the long-range dependencies among pixels [22, 23]. Inspired by U-Net Transformer [8], each stage of the U-Net's upsampling pathway is enhanced by incorporating a multi-head cross-attention (MHCA) module, which reduces noise and irrelevant elements in the skip-connected features.

For parameter-efficient transfer learning, we apply an adapter module after the feature extraction in each ViT block and scale its output using a scaling factor (empirically $s = 0.5$) to balance the task-agnostic features and the task-specific features [24]. During model fine-tuning, we freeze the pre-trained SAM weights for spatial MHSA and feed-forward layers (marked by the "snowflake" icon in Fig. 1) to leverage SAM's inherent generalizability abilities, while fine-tuning all other parts of the model including the temporal MHSA, adapter module and U-Net layers. Our model is trained to segment myocardium in a CMR slice spanning across one cardiac cycle *simultaneously*. For image preprocessing 2D+T data, the input slice is center cropped to [H, W] = [128, 128]. Image intensities were normalized to [0,1] via min-max normalization method. The model was trained using combined Dice and cross-entropy loss. The model was trained and tested on a single DGX-A100 GPU (NVIDIA, CA, USA). The GPU memory usage was around 11GB during operation and the average time to generate 2D+T segmentation for one SAX slice was 0.476 ± 0.001 s.

3 Experiments

We used a multi-center, multi-vendor cine CMR dataset (STACOM2011) to evaluate the segmentation accuracy of our data-specific model. To validate the zero-shot generalizability, we fine-tuned our model on the STACOM2011 and then applied it zero-shot on two other large datasets (ACDC and M&Ms) that were unseen during fine-tuning.

Fine-Tuning Dataset. We utilized the STACOM 2011 dataset [25], which is a multi-center, multi-vendor dataset comprising 100 patients with coronary artery disease and prior myocardial infarction. MR vendors used include GE, Philips and Siemens. Pixel-wise segmentation of the myocardium for all cardiac phases is publicly available as ground truth. Both SAX and LAX images are included. We selected this dataset since it is the only public dataset with ground truth labels available for all cardiac phases, making it suitable for training a 2D+T segmenter. In the experiments, we first split the dataset into 60:40 for training/validation and testing to evaluate segmentation accuracy of the data-specific models. Then we fine-tuned the model on the entire STACOM for zero-shot generalization evaluation on two testing datasets.

Testing Dataset. We utilized the ACDC dataset [26] and the M&Ms dataset [14] for zero-shot generalization evaluation. The ACDC dataset is a single-center, single-vendor dataset composed of 100 cine CMR cases from five different pathological groups. All images were from a single center using Siemens scanners. The M&Ms dataset is a multi-center, multi-vendor dataset composed of 136 cases in its testing cohort from 5 medical centers (4 from Spain, 1 from Germany) and 4 MR vendors (GE, Philips, Siemens and Canon). This cohort has a wide collection of cardiovascular diseases. In both datasets, the ground truth labels are available for only end-diastole (ED) and end-systole (ES) phases in SAX slices.

Comparison with SOTA Methods. We compared our method with three representative SOTA methods to address 2D+T cine CMR segmentation. DeepIED [4] (noted as "2D recurrent" in Tables 1 and 2) represents adding recurrent layers in a 2D CNN to encode sequential information along the temporal axis. nnUNet-3D [27] represents utilizing a 3D CNN with 3D kernels that treat the temporal axis as a depth dimension. U-transformer [8], which shares the same U-Net framework design as ours, represents employing attention mechanisms. We added the temporal attention to the original U-transformer to enable temporal-spatial information extraction. Note we only fine-tuned and evaluated the models on SAX images and our model was free of prompts in these comparison studies, leaving the LAX and prompts to the next section.

We first trained all methods on STACOM2011 using 60:40 split (60 training/validation, 40 testing) to evaluate the data-specific model performance. Table 1 shows the Dice coefficient and Hausdorff distance (HD). We show that our cineCMR-SAM significantly outperforms (p < 0.001 by one-tailed Wilcoxon signed-rank test) the CNN-based methods (2D recurrent and nnUNet3D) as well as the fusion of U-Net and transformer (U-transformer) for all levels of SAX slices (basal, mid and apical). We then fine-tuned all methods using the entire STACOM2011 dataset and evaluated zero-shot generalization on ACDC and M&Ms datasets. Table 2 shows that our method has significantly better generalization ability in both ACDC (Dice = 0.850 and HD = 4.066 pixels for myocardium in all SAX slices, p < 0.001 by one-tailed Wilcoxon signed-rank test) and M&Ms (Dice = 0.835 and HD = 4.674 pixels for myocardium in all SAX slices, p < 0.001) datasets compared to all other methods. The zero-shot performance on these two datasets can be found in Fig. 3. Especially, Fig. 2 shows that in M&Ms dataset there were no significant differences in our models's performance across all vendors and centers using one-way analysis of variance (ANOVA), except for vendor 1 used in center 1 with lower performance (p < 0.001 by Tukey's range test).

Notably, vendor 4 (Canon) belongs to a vendor that was not included in the fine-tuning dataset, yet our model achieved performance comparable to that with vendors seen in the fine-tuning dataset.

Table 1. Results of data-specific models on STACOM Dataset. The dataset was split 60:40 and the latter were for testing. HD = Hausdorff Distance (unit: pixel)

Methods	All SAX		Basal SAX		Mid SAX		Apical SAX	
Metrics	Dice	HD	Dice	HD	Dice	HD	Dice	HD
2D Recurrent [4]	0.865	3.842	0.877	3.214	0.874	3.126	0.813	3.811
nnUnet3D [27]	0.877	3.867	0.888	3.221	0.882	3.345	0.839	3.293
U-transformer [8]	0.880	3.622	0.890	3.044	0.887	2.874	0.836	3.486
Ours	**0.890**	**2.853**	**0.897**	**2.471**	**0.898**	**2.411**	**0.851**	**2.824**

Table 2. Results of zero-generalization on ACDC and M&Ms dataset. The models were fine-tuned on the entire STACOM dataset and directly applied to new datasets.

Methods	All SAX		Basal SAX		Mid SAX		Apical SAX	
ACDC	Dice	HD	Dice	HD	Dice	HD	Dice	HD
2D Recurrent [4]	0.809	6.720	0.825	6.344	0.811	6.181	0.757	5.917
nnUnet3D [27]	0.785	7.416	0.806	6.482	0.785	6.257	0.729	6.880
U-transformer [8]	0.828	4.741	0.846	4.321	0.829	4.092	0.790	4.264
Ours	**0.850**	**4.066**	**0.871**	**3.202**	**0.844**	**3.776**	**0.793**	**3.788**
M&Ms								
2D Recurrent [4]	0.786	6.525	0.793	7.530	0.786	7.079	0.763	6.184
nnUnet3D [27]	0.801	6.231	0.806	5.028	0.810	5.405	0.750	6.081
U-transformer [8]	0.810	5.611	0.803	6.397	0.820	5.146	0.781	4.732
Ours	**0.835**	**4.674**	**0.839**	**4.045**	**0.844**	**3.974**	**0.797**	**4.498**

Table 3. Effectiveness of prompt features. We fine-tuned both models (cineCMR-SAM without or with text prompts) on the first 60% of STACOM and tested on the rest 40%.

Prompts	All SAX		All LAX	
	Dice	HD	Dice	HD
without text	0.881	3.133	0.837	6.264
with text	**0.890**	**2.909**	**0.845**	**5.440**

Effectiveness of Text Prompts. In this study, we enabled the text prompt feature and assessed its effectiveness to improve segmentation across all types of CMR views. Concretely, we evaluated two versions of cineCMR-SAM: one with text prompts and the other without. We split STACOM2011 into a 60:40 ratio and fine-tuned the models on both SAX and LAX images together. Table 3 shows that enabling text prompts in our model increases the Dice coefficient from 0.881 to 0.890 in SAX and from 0.837 to 0.845 in LAX, while decreases the HD from 3.133 pixels to 2.909 in SAX and from 6.264 to 5.440 in LAX. These results indicate that specifying the view type of input slices enhances segmentation across all views. Notably, compared to the values reported in the section above (i.e., cineCMR-SAM without prompts trained only on SAX images), cineCMR-SAM without prompts trained on both view types shows a slight drop in performance in SAX (Dice drops 0.890 to 0.881), likely due to the model needing to learn from different input types. After using prompts, the performance returns to the same level ($p = 0.167$ by two-tailed Wilcoxon signed rank test).

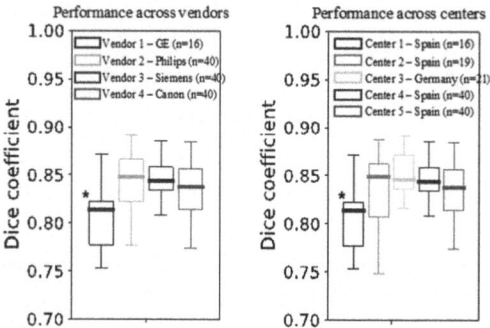

Fig. 2. The zero-shot generalization of our model across different vendors and centers in M&Ms dataset. The data are from 4 vendors and 5 centers, where center 1 uses vendor 1, center 2 and 3 uses vendor 2, center 4 uses vendor 3 and center 5 uses vendor 4. The "*" represent that Dice in vendor 1/center1 is statistically lower compared to the others ($p < 0.001$ by Tukey's range test).

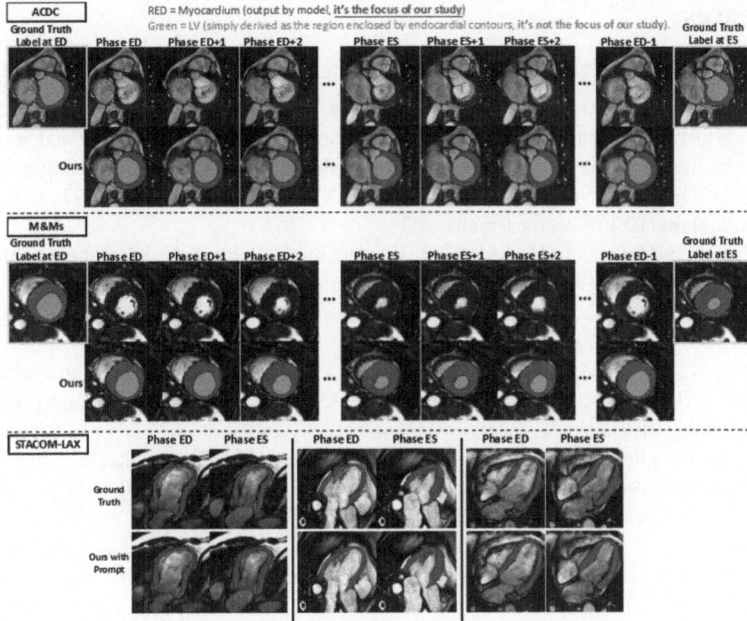

Fig. 3. The examples of our model's zero-shot performance. Ground truth labels at ED and ES for ACDC and M&Ms examples are shown in yellow box. Note that green represents left ventricle (LV) which is put here only for illustration. Our study focuses on the myocardium.

4   Conclusion

In this study, we propose cineCMR-SAM which employs SAM ViT blocks with pre-trained SAM weights and temporal-spatial adaptation. This model delivers accurate and consistent segmentation of myocardium throughout all cardiac phases within a single cardiac cycle in cine CMR scans, outperforming SOTA methods both in data-specific models and in zero-shot generalization on unseen datasets. Incorporating text prompts which specifies the input view type, our model adeptly manages segmentation across both SAX and LAX views, laying groundwork for future adaptability to diverse inputs. More detailed text prompts such as "Basal SAX" or "three-chamber LAX" should be investigated in the future. We also plan to add regularization in training loss based on known cardiac behaviors such as smooth temporal motion and decreasing volume size during systole [9] in the future work. The validation study using multi-center in-house clinical CMR datasets and assessment of the clinical parameters derived from the segmentation results is currently ongoing.

References

1. Zeppenfeld, K., et al.: 2022 ESC Guidelines for the management of patients with ventricular arrhythmias and the prevention of sudden cardiac death: developed by the task force for the management of patients with ventricular arrhythmias and the prevention of sudden cardiac death of the European Society of Cardiology (ESC) Endorsed by the Association for European Paediatric and Congenital Cardiology (AEPC). Eur. Heart J. **43**(40), 3997–4126 (2022). https://doi.org/10.1093/eurheartj/ehac262

2. Sardanelli, F., Quarenghi, M., Di Leo, G., Boccaccini, L., Schiavi, A.: Segmentation of cardiac cine MR images of left and right ventricles: Interactive semiautomated methods and manual contouring by two readers with different education and experience. J. Magn. Reson. Imaging **27**(4), 785–792 (2008). https://doi.org/10.1002/jmri.21292

3. Savioli, N., Vieira, M.S., Lamata, P., Montana, G.: Automated segmentation on the entire cardiac cycle using a deep learning work-flow. arXiv, Aug. 31, 2018. https://doi.org/10.48550/arXiv.1809.01015

4. "Cardiac-DeepIED: automatic pixel-level deep segmentation for cardiac bi-ventricle using improved end-to-end encoder-decoder network. IEEE J. Transl. Eng. Health Med. **7**, 1900110 (2019). https://doi.org/10.1109/JTEHM.2019.2900628

5. Vesal, S., Gu, M., Maier, A., Ravikumar, N.: Spatio-temporal multi-task learning for cardiac MRI left ventricle quantification. IEEE J. Biomed. Health Inform. **25**(7), 2698–2709 (2021). https://doi.org/10.1109/JBHI.2020.3046449

6. Patravali, J., Jain, S., Chilamkurthy, S.: 2D-3D fully convolutional neural networks for cardiac MR segmentation. arXiv, Jul. 31, 2017. https://doi.org/10.48550/arXiv.1707.09813

7. Bertasius, G., Wang, H., Torresani, L.: Is space-time attention all you need for video understanding? arXiv, Jun. 09, 2021. https://doi.org/10.48550/arXiv.2102.05095

8. Petit, O., Thome, N., Rambour, C., Soler, L.: U-Net transformer: self and cross attention for medical image segmentation. arXiv, Mar. 12, 2021. https://doi.org/10.48550/arXiv.2103.06104

9. Qi, X., He, Y., Qi, Y., Kong, Y., Yang, G., Li, S.: STANet: spatio-temporal adaptive network and clinical prior embedding learning for 3D+T CMR segmentation. IEEE J. Biomed. Health Inform. **28**(2), 881–892 (2024). https://doi.org/10.1109/JBHI.2023.3337521

10. Myronenko, A., et al.: 4D CNN for semantic segmentation of cardiac volumetric sequences. In: Pop, M., et al. (eds.) Statistical Atlases and Computational Models of the Heart. Multi-Sequence CMR Segmentation, CRT-EPiggy and LV Full Quantification Challenges, LNCS, pp. 72–80. Springer, Cham (2020). https://doi.org/10.1007/978-3-030-39074-7_8

11. Chen, Z., Ren, H., Li, Q., Li, X.: Motion correction and super-resolution for multi-slice cardiac magnetic resonance imaging via an end-to-end deep learning approach. Comput. Med. Imaging Graph. **115**, 102389 (2024). https://doi.org/10.1016/j.compmedimag.2024.102389

12. Bai, W., et al.: Automated cardiovascular magnetic resonance image analysis with fully convolutional networks. J. Cardiovasc. Magn. Reson. **20**(1), 65 (2018). https://doi.org/10.1186/s12968-018-0471-x

13. Chen, C., et al.: Improving the generalizability of convolutional neural network-based segmentation on CMR images. Front. Cardiovasc. Med. **7** (2020). https://doi.org/10.3389/fcvm.2020.00105

14. Campello, V.M., et al.: Multi-centre, multi-vendor and multi-disease cardiac segmentation: the M&Ms challenge. IEEE Trans. Med. Imaging **40**(12), 3543–3554 (2021). https://doi.org/10.1109/TMI.2021.3090082

15. Chen, C., et al.: Deep learning for cardiac image segmentation: a review. Front. Cardiovasc. Med. **7** (2020). https://doi.org/10.3389/fcvm.2020.00025

16. Kirillov, A., et al.: Segment anything. arXiv, Apr. 05, 2023. https://doi.org/10.48550/arXiv.2304.02643

17. Ma, J., He, Y., Li, F., Han, L., You, C., Wang, B.: Segment anything in medical images. Nat. Commun. **15**(1), 654 (2024). https://doi.org/10.1038/s41467-024-44824-z

18. Chen, C., et al.: MA-SAM: modality-agnostic SAM adaptation for 3D medical image segmentation. arXiv, Sep. 15, 2023. https://doi.org/10.48550/arXiv.2309.08842

19. Zhang, Y., Shen, Z., Jiao, R.: Segment anything model for medical image segmentation: current applications and future directions. Comput. Biol. Med. **171**, 108238 (2024). https://doi.org/10.1016/j.compbiomed.2024.108238

20. Kim, S., et al.: MediViSTA-SAM: zero-shot medical video analysis with spatio-temporal SAM adaptation. arXiv, Nov. 13, 2023. http://arxiv.org/abs/2309.13539. Accessed 10 Jan 2024

21. Shaker, M.S., Wael, M., Yassine, I.A., Fahmy, A.S.: Cardiac MRI view classification using autoencoder. In: 2014 Cairo International Biomedical Engineering Conference (CIBEC), pp. 125–128 (2014). https://doi.org/10.1109/CIBEC.2014.7020935

22. Chen, J., et al.: TransUNet: transformers make strong encoders for medical image segmentation, arXiv:2102.04306 [cs], Feb. 2021. http://arxiv.org/abs/2102.04306. Accessed 22 Mar 2022

23. Wang, H., Cao, P., Wang, J., Zaiane, O.R.: UCTransNet: rethinking the skip connections in U-Net from a channel-wise perspective with transformer. AAAI **36**(3), 2441–2449 (2022). https://doi.org/10.1609/aaai.v36i3.20144

24. Chen, S., et al.: AdaptFormer: adapting vision transformers for scalable visual recognition. arXiv, Oct 14, 2022. https://doi.org/10.48550/arXiv.2205.13535

25. Suinesiaputra, A., et al.: A collaborative resource to build consensus for automated left ventricular segmentation of cardiac MR images. Med. Image Anal. **18**(1), 50–62 (2014). https://doi.org/10.1016/j.media.2013.09.001

26. Bernard, O., et al.: Deep learning techniques for automatic MRI cardiac multi-structures segmentation and diagnosis: is the problem solved? IEEE Trans. Med. Imaging **37**(11), 2514–2525 (2018). https://doi.org/10.1109/TMI.2018.2837502

27. Isensee, F., Jaeger, P.F., Kohl, S.A.A., Petersen, J., Maier-Hein, K.H.: NnU-Net: a self-configuring method for deep learning-based biomedical image segmentation. Nat. Methods **18**(2), 203–211 (2021). https://doi.org/10.1038/s41592-020-01008-z

Navigating Data Scarcity Using Foundation Models: A Benchmark of Few-Shot and Zero-Shot Learning Approaches in Medical Imaging

Stefano Woerner[1]([✉]) and Christian F. Baumgartner[1,2]

[1] Cluster of Excellence "Machine Learning", University of Tübingen, Tübingen, Germany
{stefano.woerner,christian.baumgartner}@uni-tuebingen.de
[2] Faculty of Health Sciences and Medicine, University of Lucerne, Lucerne, Switzerland

Abstract. Data scarcity is a major limiting factor for applying modern machine learning techniques to clinical tasks. Although sufficient data exists for some well-studied medical tasks, there remains a long tail of clinically relevant tasks with poor data availability. Recently, numerous foundation models have demonstrated high suitability for few-shot learning (FSL) and zero-shot learning (ZSL), potentially making them more accessible to practitioners. However, it remains unclear which foundation model performs best on FSL medical image analysis tasks and what the optimal methods are for learning from limited data. We conducted a comprehensive benchmark study of ZSL and FSL using 16 pretrained foundation models on 19 diverse medical imaging datasets. Our results indicate that BiomedCLIP, a model pretrained exclusively on medical data, performs best on average for very small training set sizes, while very large CLIP models pretrained on LAION-2B perform best with slightly more training samples. However, simply fine-tuning a ResNet-18 pretrained on ImageNet performs similarly with more than five training examples per class. Our findings also highlight the need for further research on foundation models specifically tailored for medical applications and the collection of more datasets to train these models.

1 Introduction

Machine learning is revolutionizing the field of medical imaging and diagnostics, offering capabilities that were previously unattainable. However, these advancements typically depend on the availability of large, well-annotated datasets. For many medical applications, such as the diagnosis of rare diseases, collecting these types of datasets is often infeasible. Consequently, in many real-world scenarios,

Supplementary Information The online version contains supplementary material available at https://doi.org/10.1007/978-3-031-73471-7_4.

Z. Deng et al. (Eds.): MedAGI 2024, LNCS 15184, pp. 30–39, 2025.
https://doi.org/10.1007/978-3-031-73471-7_4

there is often insufficient data to effectively train highly performant deep learning models. Additionally, computational resources are frequently limited, which poses further challenges in training or even fine-tuning state-of-the-art models.

Few-shot learning (FSL) has shown great potential in addressing these data-scarce applications. With effective FSL strategies, clinics and medical researchers could potentially train models using their own small datasets and achieve performance levels acceptable for clinical practice. Few-shot learning is most commonly performed through fine-tuning of a large pretrained model on the smaller, domain-specific, target dataset. Recently, several large models, known as foundation models, have been published after being trained on vast amounts of data [2,8,9,16]. Many such models have been shown to have excellent generalization capabilities, and to be highly suitable for FSL. However, no large-scale studies exist which compare FSL performance of different pretrained models across a broad and diverse array of medical imaging domains. A number of foundation models are also capable of zero-shot learning (ZSL) by searching for the highest correspondence between the representations of the input image and a language prompt. Similarly, there are no works rigorously comparing the ZSL capabilities of different foundation models on a diverse range of medical tasks.

In this paper, we present the first large-scale study comparing the FSL and ZSL performance of various publicly available pretrained models across a diverse set of medical imaging domains. We conduct our study on the recently released MedIMeta dataset [15], which is comprised of 19 different datasets from 10 different imaging modalities and anatomical regions. In comprehensive experiments we evaluate 16 publicly available models that have been pretrained on different medical and non-medical data sources. Because fine-tuning very large models is not practical within the computational budget of most clinicians and researchers, we limited ourselves to exploring strategies that are possible to perform in the realistic scenario of having access to a single mid-range to high-end GPU. Within these constraints we explore a linear probing strategy as well as fine-tuning. For the five models in our benchmark that support ZSL, we also benchmark their ZSL capabilities with different prompt styles.

Our experiments yield a number of practical insights and actionable recommendations. We make code to reproduce our results and adapt our experiments publicly available.[1]

2 Methods

2.1 Dataset

To allow us to study the FSL and ZSL performance on wide array of different image modalities and tasks, we conduct our experiments on the recently released MedIMeta dataset [14,15]. MedIMeta is a highly standardized meta-dataset compiled from 19 publicly available datasets, and covering 10 different imaging modalities. We use the main (i.e. first) task for each of the 19 datasets.

---

[1] https://github.com/StefanoWoerner/medimeta-fsl-benchmark.

We refer the reader to [14,15] for detailed descriptions of the sub-datasets and tasks.

2.2   Simulation of FSL and ZSL Tasks

We artificially construct multiple FSL tasks from each of MedIMeta's datasets by randomly sampling n labeled training samples and 10 unlabeled query samples per class from each dataset. We ensure that no images of the same subject are spread over the two sets. We coin these individual FSL tasks a *task instance*. To ensure robust FSL performance measurement, we randomly generate 100 task instances for each dataset and average the results. In order to investigate the effect of increasing numbers of labeled training samples we repeat all experiments for $n \in \{1, 2, 3, 5, 7, 10, 15, 20, 25, 30\}$. In addition we also simulate task instances with $n = 0$, i.e. only query samples, for the ZSL evaluation.

2.3   Pretrained Models

We evaluate three distinct pretraining paradigms: supervised pretraining, self-supervised pretraining, and contrastive language-image pretraining (CLIP). In the following we briefly describe the specific architectures and pretraining data.

Fully Supervised Models. We investigate the widely used **Residual Networks (ResNet) architecture** [6] in the variations ResNet18, ResNet50, and Resnet101, all of which have been pretrained on the ImageNet dataset [11].

We further investigate the **Vision Transformer (ViT) architecture** [4]. Due to it's excellent performance on many computer vision benchmarks, the ViT has become a standard architecture and the basis of a large amount of further work. We compare the base (ViT-B), large (ViT-L), and huge (ViT-H) architecture variations with patch sizes 16, 16 and 14, respectively. We consider models pretrained on ImageNet [11] and on ImageNet21k [10].

Self-supervised Models. In this category we consider the **self-DIstillation with NO labels (DINO)** model [1]. We specifically focus on the recently released DINOv2 model [8] which relies on a ViT architecture that was pretrained using a self-supervised knowledge distillation approach. The model was trained using a very large unlabeled but curated dataset assembled from various computer vision datasets. The DINOv2 representations have been shown to be highly transferable across computer vision tasks [8]. We consider the ViT-B, ViT-L, and giant (ViT-g) variations with patch size 14.

Contrastive Language-Image Pretraining. Lastly, we consider two CLIP models which employ contrastive learning to align images and text into a shared embedding space [9].

Firstly, we use the original **CLIP model** with the weights for ViT-B and ViT-L provided by OpenAI [9]. These models have been pretrained on 400 million image-text pairs collected from the internet. Although the specific composition of this dataset is not traceable, it is likely that a small portion of medical data was included. In addition to its unique ZSL capabilities, the CLIP model was also shown to perform extremely well on computer vision FSL tasks by training a linear probe on the final image-encoder representations [9].

Secondly, we use the ViT-H and ViT-g models trained on LAION-2B [12] provided by OpenCLIP [2], an open source reimplementation of OpenAI's CLIP. LAION-2B contains 2 billion image-text pairs extracted from common crawl [3] and is the English language subset of the larger LAION-5B [12] dataset. Similar to the OpenAI data, the inclusion of small amounts of medical data is likely.

We also investigate the **BiomedCLIP model** [16] which uses the same ViT architecture as the base version of the original CLIP, but replaces the text encoder with PubMedBERT [5], a language model tailored for the biomedical domain. BiomedCLIP was pretrained on 15 million text-image pairs extracted from PubMed articles (PMC-15M). This is the only model in our study that was trained exclusively on medical data. BiomedCLIP can be employed for FSL and prompt-based ZSL in the same manner as CLIP.

2.4   Few-Shot Learning Strategies

We evaluate two model adaptation strategies: fine-tuning and linear probing.

Fine-tuning involves initializing a network with pretrained weights, and then continuing the training of all weights in the network with an FSL task instance. The last linear layer (classification layer) is replaced with a new layer matching the number of classes in the target task. For most foundation models, which commonly have hundreds of millions or even billions of parameters, fine-tuning is computationally infeasible for many practitioners. We therefore only evaluate the fine-tuning strategy on the ResNet-18 and ResNet-50 variants.

Similarly, **linear probing** involves initializing a network with pretrained weights, and attaching a new classification layer. However, in linear probing the backbone network is frozen, and a simple linear classifier is trained on the final representations of the network. This was shown to lead to strong FSL performance assuming the base network is able to extract useful image features [9]. Since only a linear classifier is trained on the image features produced by the pretrained network, this strategy is computationally much cheaper than fine-tuning the complete network, making it feasible to use with large foundation models.

We conduct an extensive **hyper-parameter search** on a separate set of sampled FSL tasks for both fine-tuning and linear probing. For each of the models and each number of labeled samples n, we test two optimizers (SGD and Adam), two different head initialization strategies (Kaiming initialization [6], initialization with all zeros [13]), a range of learning rates between 10^{-5} and 0.1, and a range of training steps between 5 and 200. We evaluate all models from Sect. 2.3 using their respective optimal parameters from the hyper-parameter search.

2.5   Zero-Shot Learning Strategy

The CLIP [9] and BiomedCLIP [16] models have the capability of solving classification tasks with no labeled training examples by searching for the highest similarity between an input image and several text prompts corresponding to different target classes. We test three different prompt templates. First, we investigate simply using the class names extracted from the MedIMeta task definitions as prompts. Secondly, we test two templates which add information about the imaging modality: "A {domain_identifier} image where the {task_name} is {class_name}", and "This {domain_identifier} image shows [a] {class_name}". All variables above are extracted from the MedIMeta task description files. However, some class names and domain identifiers needed to be adjusted in order to form a grammatically correct and semantically meaningful sentences.

2.6   Metrics

We evaluate the performance for each dataset and each training set size n using the area under the receiver operator curve (AUROC) averaged over all 100 task instances. To obtain a measure of average performance across all datasets, we use the harmonic mean of the AUROCs from each dataset.

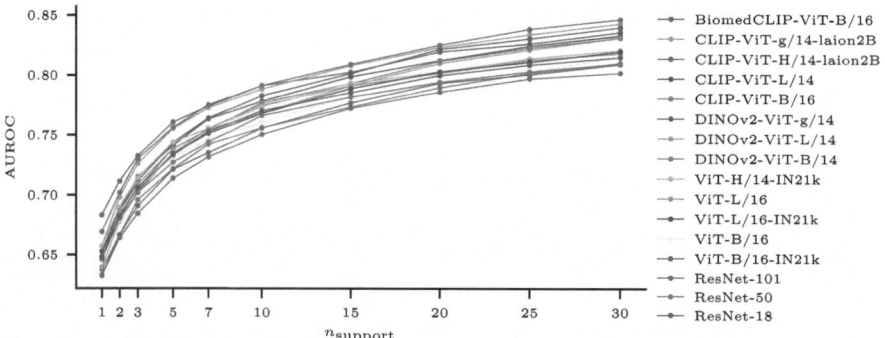

Fig. 1. Harmonic mean AUROC over all 19 MedIMeta datasets

3   Experiments and Results

We performed all FSL and ZSL experiments using all models and learning strategies as described above. In the following, we describe our main findings. All results can be found in Table A.1 in the Supplementary Material.

The Optimal Hyperparameters were Similar for All Models. For all models the best-performing optimizer was Adam [7]. Further, initializing the classification head with zeros performed better or on par compared to Kaiming

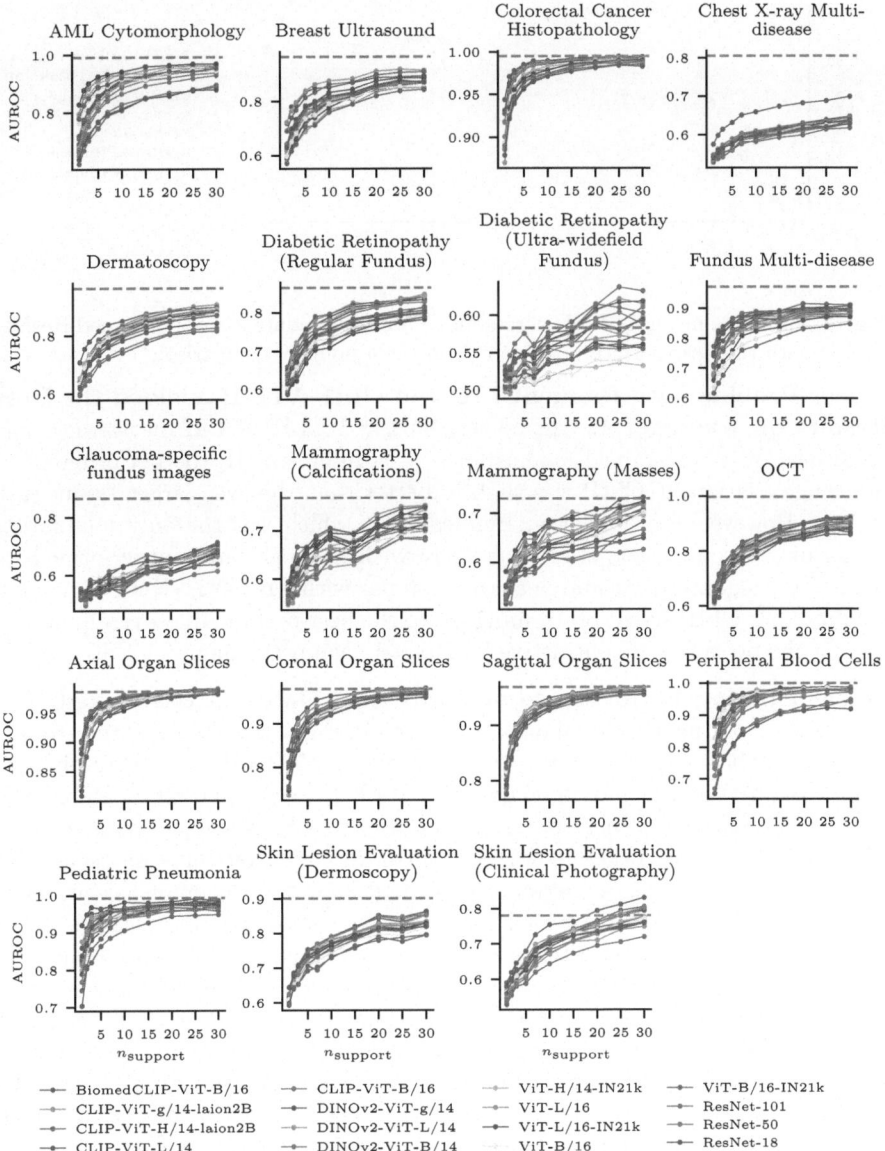

Fig. 2. AUROC on the different datasets with fully supervised baseline from [15]. The fully-supervised performance is indicated by the black dotted line.

initialization [6], in line with the findings in [13]. For most models and n, using a learning rate of 10^{-4} with at least 120 training steps was optimal or close to optimal.

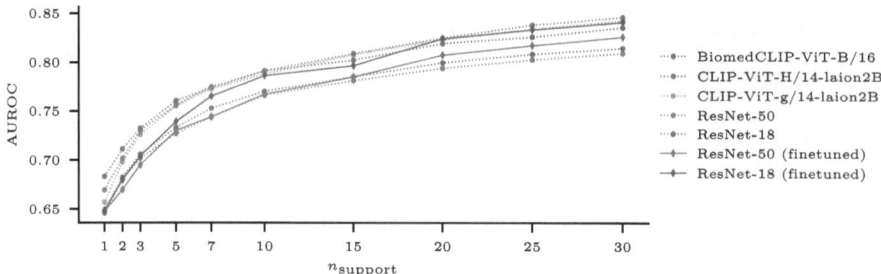

Fig. 3. Harmonic mean AUROC across all 19 MedIMeta datasets of fine-tuned ResNet models with the best-performing linear probe as a point of comparison.

Linear Probing with BiomedCLIP and CLIP-ViT-H Yielded the Best Results on Average. As can be seen in Fig. 1 CLIP-ViT-H on average outperformed other pretrained models for $n \geq 7$. Interestingly, the performance of the "huge" variant of CLIP was slightly better than the even larger "giant (g)" variant. However, for smaller n, BiomedCLIP, which was the only foundation model in our comparison trained entirely with medical data, outperformed its larger CLIP counterparts and performed on par with CLIP-ViT-H up to $n = 10$. We note that CLIP-based pretraining led to the best performance overall, underscoring the potential of contrastive language-image pretraining for FSL.

Linear Probing Performance on Individual Datasets was Mixed. The performance on the individual datasets shown in Fig. 2 was mixed. Interestingly, the method that performed best on average for $n \geq 7$ (CLIP-ViT-H) rarely performed the best on the individual datasets. Rather it was consistently among the top-few approaches on most datasets leading to its high average performance. BiomedCLIP, on the other hand, performed very well on some datasets (e.g. Chest X-ray Multi-disease, Dermatoscopy, and Pediatric Pneumonia), but more poorly on others (e.g. Ultra-widefield Fundus, or Mammography (Masses)). We hypothesize that BiomedCLIP performed more strongly on images that were overrepresented in the PMC-15M pretraining dataset. We conclude that in practice linear probing on the BiomedCLIP model might often be a good first attempt when working with very few labeled images, but it does not obviate thorough evaluation on a held-out test set. With more training data, linear probing CLIP-ViT-H is likely the better option, especially since it does not display as much variablility throughout different imaging modalities as BiomedCLIP.

FSL Performed Close to Fully Supervised Learning for Some Tasks. In Fig. 2 we additionally show the fully supervised baseline performance on the official data splits reported in [14]. We observed that for some tasks the 30-shot performance almost matched the fully supervised performance. Indeed, for the Ultra-widefield Fundus dataset the FSL performance was substantially better than the fully supervised performance. We believe this is due to the small number of training images in the official split of this dataset. Nevertheless, this suggests that for some problems linear probing of a foundation model may be a better alternative than training a model from scratch with a small dataset.

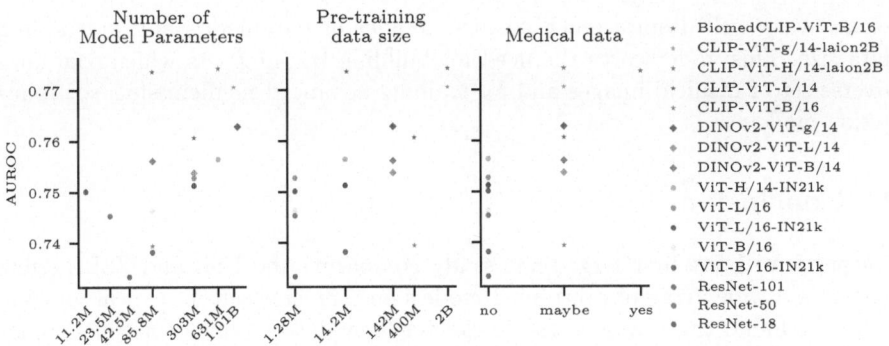

Fig. 4. Model properties plotted against mean few-shot performance over the tested n. The symbol indicates the type of pretraining (Supervised, DINO, CLIP).

Linear Probes on Large Models Beat Fine-Tuning of Small Models. In Fig. 3 it can be seen that linear probing with CLIP-ViT-H on average outperformed fine-tuning of the ResNet-18 and ResNet-50 for all n. However, with more training data fine-tuning the ResNet-18 performed *almost* as well as CLIP-ViT-H, and for $n \geq 20$ the fine-tuned ResNet-18 outperformed BiomedCLIP. Interestingly, fine-tuning ResNet-18 clearly outperformed fine-tuning ResNet-50, suggesting that a lower network complexity may be preferable in the FSL scenario. Our findings suggest that while linear probing very large foundation models such as the CLIP-ViT-H on average may lead to small performance gains, the commonly used strategy of fine-tuning a ResNet-18 also performs strongly given sufficient data. We note that while fine-tuning of foundation models may lead to even better results, this is computationally prohibitive for the majority of practitioners.

ZSL Performance Could not Match FSL Performance. The ZSL approaches had average AUROC scores ranging from 0.316 to 0.397, far below those of the 1-shot performance reported in Fig. 1 where average AUROCs range from 0.632 to 0.683. This contradicts the findings of Radford et al. [9] who showed that on computer vision tasks the CLIP model can often outperform linear probes in the ZSL setting. We conclude that ZSL may not yet be a suitable strategy for general medical image analysis tasks. We report the ZSL results in Fig. A.1 in the Supplementary Materials.

Model Complexity and Pretraining Data Size Correlate with Performance. In Fig. 4, we explore the relation of the following three model properties to their linear probing performance: the model size, the number of samples in the pretraining data, and the type of pretraining data. We observed that there was a strong positive correlation between model size and few-shot performance as well as pretraining set size and few-shot performance. While the non-medical CLIP-ViT-H, and CLIP-ViT-g clearly outperformed all other non-medical approaches on average, BiomedCLIP, which was trained on medical data exclusively, performed very well despite much smaller number of parameters and pretraining data size. This underscores the need for building training sets which contain a diverse set of medical images and for training advanced medicine-focused foundation models.

4   Conclusion

We performed the first large-scale study comparing the FSL and ZSL performance of a wide array of pretrained models on a diverse set of medical imaging data. We found that, on average, in the very low data regime of $n \leq 5$ samples per class, a linear probe on BiomedCLIP was the best strategy. However, with more data, linear probing of the CLIP-ViT-H model performed slightly better. While fine-tuning a ResNet-18 on average performed worse compared to a linear probe on CLIP-ViT-H, it still reached a high performance for $n \geq 20$. We also observed a large variance between the performance on the different datasets emphasizing the need for cautious application of these technologies. Our investigation further revealed that parameter-rich foundation models trained on very large non-medical datasets have very good FSL performance on medical tasks. However, the strong performance of BiomedCLIP model on some datasets underscores the potential of foundation models specific to medical applications.

Acknowledgments. Funded by the Deutsche Forschungsgemeinschaft (DFG, German Research Foundation) under Germany's Excellence Strategy - EXC number 2064/1 - Project number 390727645. The authors thank the International Max Planck Research School for Intelligent Systems (IMPRS-IS) for supporting Stefano Woerner.

Disclosure of Interests. The authors have no competing interests to declare.

References

1. Caron, M., et al.: Emerging properties in self-supervised vision transformers. In: Proceedings of the IEEE/CVF International Conference on Computer Vision, pp. 9650–9660 (2021)
2. Cherti, M., et al.: Reproducible scaling laws for contrastive language- image learning. In: 2023 IEEE/CVF Conference on Computer Vision and Pattern Recognition (CVPR). IEEE (2023). https://doi.org/10.1109/CVPR52729.2023.00276
3. Common Crawl. https://commoncrawl.org
4. Dosovitskiy, A., et al.: An image is worth 16×16 words: transformers for image recognition at scale. arXiv: 2010.11929 (2020)
5. Gu, Y., et al.: Domain-specific language model pretraining for biomedical natural language processing. In: ACM Transactions on Computing for Healthcare 3.1, pp. 1–23 (2021). ISSN: 2637-8051. https://doi.org/10.1145/3458754.
6. He, K., Zhang, X., Ren, S., Sun, J.: Deep residual learning for image recognition. In: Proceedings of the IEEE Conference on Computer Vision and Pattern Recognition (CVPR) (2016)
7. Kinga, D., Adam, J.B.: Adam: a method for stochastic optimization. arXiv: 1412.6980 (2017)

8. Oquab, M., et al.: Dinov2: learning robust visual features without supervision. In: arXiv preprint arXiv:2304.07193 (2023)

9. Radford, A., et al.: Learning transferable visual models from natural language supervision. In: Meila, M., Zhang, T. (eds.) Proceedings of the 38th International Conference on Machine Learning, vol. 139. Proceedings of Machine Learning Research. PMLR, 18-24 Jul 2021, pp. 8748–8763. https://proceedings.mlr.press/v139/radford21a.html

10. Ridnik, T., Ben-Baruch, E., Noy, A., Zelnik-Manor, L.: ImageNet- 21K pretraining for the masses. arXiv: 2104.10972 (2021)

11. Russakovsky, O., et al.: ImageNet large scale visual recognition challenge. Int. J. Comput. Vis. (IJCV) **115**(3), 211–252 (2015). https://doi.org/10.1007/s11263-015-0816-y.

12. Schuhmann, C., et al.: LAION-5B: an open large-scale dataset for training next generation image-text models. arXiv: 2210.08402 (2022)

13. Woerner, S., Baumgartner, C.F.: Strategies for meta- learning with diverse tasks. In: Medical Imaging with Deep Learning (2022)

14. Stefano Woerner, S., Jaques, A., Baumgartner, C.F.: MedIMeta: A comprehensive and easy-to-use multi- domain multi-task medical imaging meta-dataset (2024). https://doi.org/10.5281/zenodo.7884735.

15. Woerner, S., Jaques, A., Baumgartner, C.F.: A comprehensive and easy-to-use multi-domain multi-task medical imaging metadataset (MedIMeta) arXiv: 2404.16000 (2024)

16. Zhang, S., et al. BiomedCLIP: a multimodal biomedical foundation model pretrained from fifteen million scientific image-text pairs. arXiv: 2303.00915 (2023)

AutoEncoder-Based Feature Transformation with Multiple Foundation Models in Computational Pathology

Woojin Chung[1], Yujun Park[2], and Yonnho Nam[1(✉)]

[1] Deparment of Biomedical Engineering, Hankuk University of Foreign Studies,
Yongin-si 17035, Gyeonggi-do, Korea
{goglxych97,yoonhonam}@hufs.ac.kr
[2] Department of Pathology, CHA Bundang Medical Center, CHA University,
Seongnam-si 13496, Gyeonggi-do, Korea
isutar_star@naver.com

Abstract. The performance of deep learning models is highly dataset-dependent. Pretrained models on large-scale datasets have significant advantages in understanding general patterns by leveraging large volumes of data. Some of these models, which are adaptable to a wide range of downstream tasks, are referred to as foundation models. Recently, several foundation models have been published in the field of computational pathology, recognized for their potential to advance deep learning applications in several downstream tasks. In order to effectively utilize multiple foundation models, each with its own advantages, it is crucial to effectively summarize or ensemble their advantages. In this paper, we propose a feature transformation method for the effective utilization of features from multiple foundation models using an autoencoder-based architecture. This method facilitates the extraction of integrated features from multiple foundation models, enabling more generalized training. We demonstrated that the proposed approach resulted in more robust representations for out-of-distribution datasets in our patch-level classification tasks.

Keywords: Computational Pathology · Foundation Model · Feature Extraction · Deep Learning · AutoEncoder

1 Introduction

The performance of deep learning models highly depends on the training datasets. Models pretrained on large-scale, high-quality datasets not only provide a strong starting point but also help in understanding general patterns, aiding in finding optimal convergence points for various tasks. These models, depending on the data spectrum they were trained on and their intended use, are called foundation models [17]. In medical imaging, where data access is limited due to patient privacy concerns and high data collection costs, foundation models

Z. Deng et al. (Eds.): MedAGI 2024, LNCS 15184, pp. 40–49, 2025.
https://doi.org/10.1007/978-3-031-73471-7_5

can be effectively utilized. They leverage pretrained information, minimizing the need for direct data access, reducing data acquisition costs, and safeguarding patient privacy. This highlights the growing importance of foundation models in the medical domain [6, 17].

Whole Slide Images (WSIs), which are often gigapixel in size, are difficult to handle as input for image processing models. Additionally, generating annotations for these images is expensive and labor-intensive work. Foundation models are considered to address these conventional challenges, and several foundation models have recently been published in the field of computational pathology [18]. The performance of these models have demonstrated on multiple benchmark datasets, proving their robustness and generalizability [7, 8]. With advancements in technology and resources, the prevalence of these models is expected to continue increasing. Therefore, the strategic use of foundation models should be investigated to advance computational pathology.

Foundation models have been trained on different datasets using various methodologies, resulting in unique strengths for each. To leverage these strengths, it is essential to have a method that can effectively summarize or ensemble these advantages. This approach is particularly useful in fields like computational pathology, where capturing the vast array of universal patterns is challenging. By using multiple foundation models information, it is possible to achieve more robust and general representations. This can lead to more accurate and reliable support diagnostic tools.

However, obtaining integrated information from multiple foundation models in computational pathology involves several considerations. First, handling the massive size of WSIs for both training and inference requires significant time. As foundation models grow in size, using multiple models in parallel without proper tuning poses problems in terms of computing resources and practical application. Additionally, since these models were trained by different institutions, the resulting features have varying properties, making it difficult to determine their correlations. This diversity complicates the task of achieving optimal performance, necessitating effective integration methods. Finally, while there may be an optimal combination of models for specific datasets and tasks, manually selecting these combinations is impractical due to time and resource constraints. Therefore, developing a robust and efficient method for integrating the knowledge from multiple foundation models is necessary.

There have been various model fusion strategies [10] in the field of deep learning to effectively utilize and integrate information from multiple models, such as multi-teacher student knowledge distillation [11, 19] and model ensembling [13]. Multi-teacher student knowledge distillation involves transferring knowledge from multiple teacher models to a single student model, thereby capturing common representations and improving the robustness of the student model. Model ensembling aggregates models using various methods to achieve better performance compared to a single model. These methods aim to enhance learning ability by improving generalization and robustness through the use of multiple pretrained models.

In this paper, we propose a feature transformation method that extracts integrated representations from multiple foundation models using an autoencoder-based architecture [9]. This approach extracts transformed features highly correlated to the original features from each of these multiple models. Using this transformed information, we conducted patch-level classification tasks in pathology and compared the performance of the proposed method with those from individual foundation models, particularly on out-of-distribution data.

2   Method

2.1   Used Pretrained Models

There are several pretrained models available in computational pathology. For our method, we utilized four models: three self-supervised pretrained models, each with a different architecture [1,3,16] and one task-specific pretrained model [15]. All of these models are trained at the patch-level. Due to being trained on vast datasets using self-supervised methods, the three self-supervised models are also classified as foundation models. Table 1 provides details of the models used.

Table 1. Details of the pretrained models used for the proposed method.

Model Name	Model Architecture	Trained Method	Trained Dataset
MONAI pathology tumor detection [15]	ResNet18 [5]	Task-specific Learning	Camelyon16
Ciga et al. [3]	ResNet18 [5]	SimCLR [2]	TCGA, TUPAC16, CPTAC, SLN-Breast Camelyon16,17
CTransPath [16]	Swin-T [12]	SimCLR [2]	TCGA, PAIP
UNI [1]	ViT-L [4]	DINOv2 [14]	Private Dataset

2.2   Proposed Method for Feature Transformation

In computational pathology, foundation models pretrained at the patch-level receive patches extracted from WSIs as input and output features. Each foundation model produces unique representations in its features due to differences in training methods and model architectures. These original features, after passing through the models, differ in both the information contained in their dimensions and their shapes. To harmonize this information, we utilized the capability of autoencoders to re-arrange dimensional information in the internal latent space during the training process. By using autoencoders, features was compressed and reconstructed, effectively filtering out noise and preserving essential information. The overview of the proposed method is illustrated in Fig. 1. Our proposed method consists of the following two steps:

In the first step, we used an autoencoder-based architecture, as shown in Fig. 1(a), to train the reconstruction of features that had passed through several foundation models. The encoder of the autoencoder effectively compressed the input data into its latent space. While training autoencoders to reconstruct the original representations from each foundation model, we imposed a constraint using the cosine similarity loss function to increase the correlation at the bottlenecks formed after passing through each encoder. This training process involved applying both the reconstruction loss for the features and the cosine similarity loss in the latent space. As the cosine similarity loss converged, these encoders explored the common representations shared by all the foundation models and constructed integrated latent spaces in the bottlenecks. We expect these highly correlated spaces to have learned a more robust representation from multiple pretrained models and to perform better on out-of-distribution data than a single foundation model.

In the second step, we used the transformed features from the trained encoder to conduct downstream classification tasks. The transformed features, compared to the original features, incorporated information from other foundation models. To demonstrate the effectiveness of the proposed method, we compared the results of training on the original features provided by the foundation model with those from training on the transformed features.

2.3 Evaluation Tasks

We evaluated our method on two tasks in computational pathology: (1) predicting lymph node metastasis in early gastric cancer and (2) cancer region segmentation in WSIs. All training steps were performed at the patch-level using 224×224 pixel size patches at 10×2 magnification. Since each pretrained model requires different input normalization, the patches were transformed accordingly for each model. To confirm whether our method produces more robust representations, we included both internal and external cases for each task. All results were compared across three scenarios: scratch learning using ResNet18 [5], transfer learning with the promising foundation model named UNI, trained with the DINOv2 framework [1,14], and our method, which involves using the transformed features from the original features of this foundation model. Transfer learning was performed using linear probing with initialized layers.

For predicting lymph node metastasis in early gastric cancer, we conducted classification on patch-level metastasis using the labels from the slide-level annotation. Afterward, we observed the probability for each patch in the slides and then calculated the Area Under the ROC Curve (AUC) value for the average probability at the slide-level. Quantitative evaluations were performed on both internal and external datasets.

Gastric cancer region segmentation was trained at the patch-level. Patches were extracted from WSIs annotated at the slide-level, and a classification task was performed to determine if each patch contained a cancer region. Quantitative evaluation was performed only on the internal dataset, as gold standard labels for the external dataset were not available.

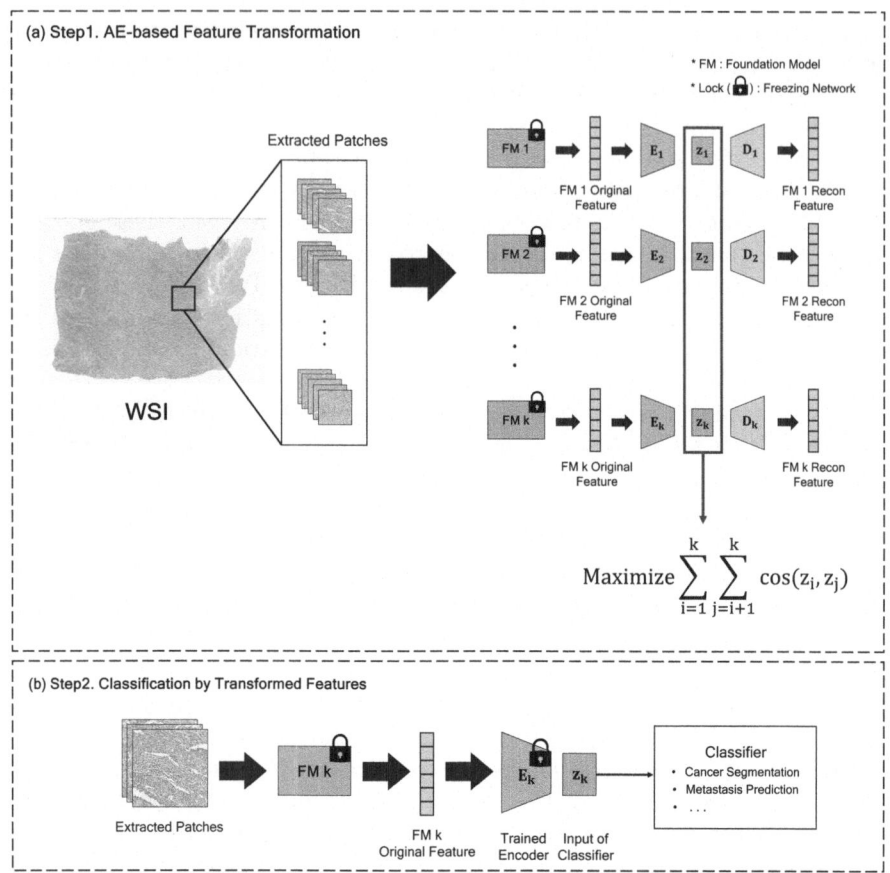

Fig. 1. Overview of the proposed method. This method consists of the following two steps. (a) Step 1: patches extracted from WSIs were used to train the reconstruction of the original features from each foundation model using autoencoders. During this process, cosine similarity loss function was employed to increase the correlation of the latent spaces at the bottleneck. (b) Step 2: after the autoencoders converged, we selected one foundation model and its corresponding encoder to conduct the classification task using the transformed features.

3   Result

3.1   Datasets

Lymph Node Metastasis in Early Gastric Cancer. We used 40×2 magnification hematoxylin and eosin stained WSIs from our institution for our study. The training dataset consisted of 200 WSIs, with 100 slides from patients with lymph node metastasis (LNM) and 100 slides from patients without LNM. These slides were sourced from patients with early gastric cancer who underwent curative surgical resection with lymph node dissection. The internal validation dataset included an additional 60 surgical cases of early gastric cancer from our institution, comprising 30 cases with LNM and 30 cases without LNM. The external validation dataset included 46 endoscopic resection cases with additional lymph node dissection from external sources, comprising 23 cases with LNM and 23 cases without LNM.

Cancer Region Segmentation. An expert pathologist annotated the cancer regions at the slide-level for 80 gastric cancer WSIs from our institution. From this dataset, 64 slides were used for training and 16 slides for internal testing. Additionally, we obtained 20 gastric cancer slides from other institution as an external dataset and performed cancer region segmentation to observe the trends in the results.

3.2 Lymph Node Metastasis Prediction in Early Gastric Cancer

We evaluated the performance of lymph node metastasis prediction in early gastric cancer for each method on both internal and external datasets. The results are shown in Fig. 2(a). For scratch learning, the AUC values for the internal and external datasets were 0.46 and 0.36, respectively, indicating overfitting due to the insufficient number of datasets, resulting in poor prediction performance. Our proposed method and the foundation model using transfer learning achieved AUC values of 0.79 and 0.77 on the internal dataset, respectively. However, on the external dataset, there was a significant difference, with AUC values of 0.71 for our method and 0.63 for the transfer learning.

3.3 Cancer Region Segmentation

The results of cancer region segmentation for the internal dataset are shown in Fig. 2(b). Both the transfer learning with foundation model and our proposed method outperformed scratch learning, with accuracy scores of 0.984

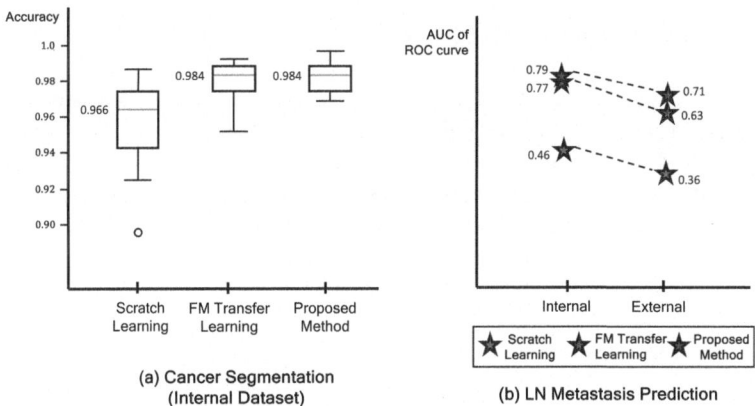

(a) Cancer Segmentation
(Internal Dataset)

(b) LN Metastasis Prediction

Fig. 2. (a) AUC values for lymph node metastasis prediction in early gastric cancer using internal and external datasets. (b) Box plots of accuracy for cancer segmentation in the internal dataset.

and 0.984 compared to 0.966, respectively. Additionally, our proposed method demonstrated more robust outcomes with less variance.

We examined the segmentation results for each patch in both the internal and external datasets, with examples shown in Fig. 3. The results from the internal dataset showed no significant difference between our proposed method and the foundation model with transfer learning. However, compared to both methods, scratch learning exhibited several false positive regions. In the external dataset, the trained scratch learning model failed to predict most of the regions. When comparing the results with and without feature transformation, the predictions made using the original features showed some false positive regions in cancer detection. As shown in Fig. 3(b), the results of transfer learning without our feature transformation method predicted muscle tissue and lymphocyte patches as cancer. Additionally, we observed that our method improved the detection of patterns in the boundary areas of WSIs and ulcer patches in our out-of-distribution segmentation.

3.4   Ablation Study on Proposed Method

An ablation study was conducted to determine whether the improved generality of the proposed method is merely attributable to the large model size. In the experiment for lymph node metastasis prediction in early gastric cancer, the same deep layers used for feature transformation and the downstream task were applied to the original features, and results were observed for each pretrained model. Evaluated based on the mean probability at the slide level, the AUC values showed that the method produced more robust representations on the external dataset compared to when the feature transformation was excluded. The AUC values increased across all foundational models, suggesting that the improved generalization is due not only to the model size but also to the effectiveness of the feature transformation approach. The results are shown in Table 2.

4   Discussion and Conclusion

In this paper, we propose a method to capture the common features of multiple foundation models. By leveraging information from several foundation models, our method can help find a more integrated representation, providing more robust results on both internal and external datasets. Processing giga-pixel sized WSIs, our method did not significantly increase the inference time compared to using a single model, as it only used one foundation model and its corresponding

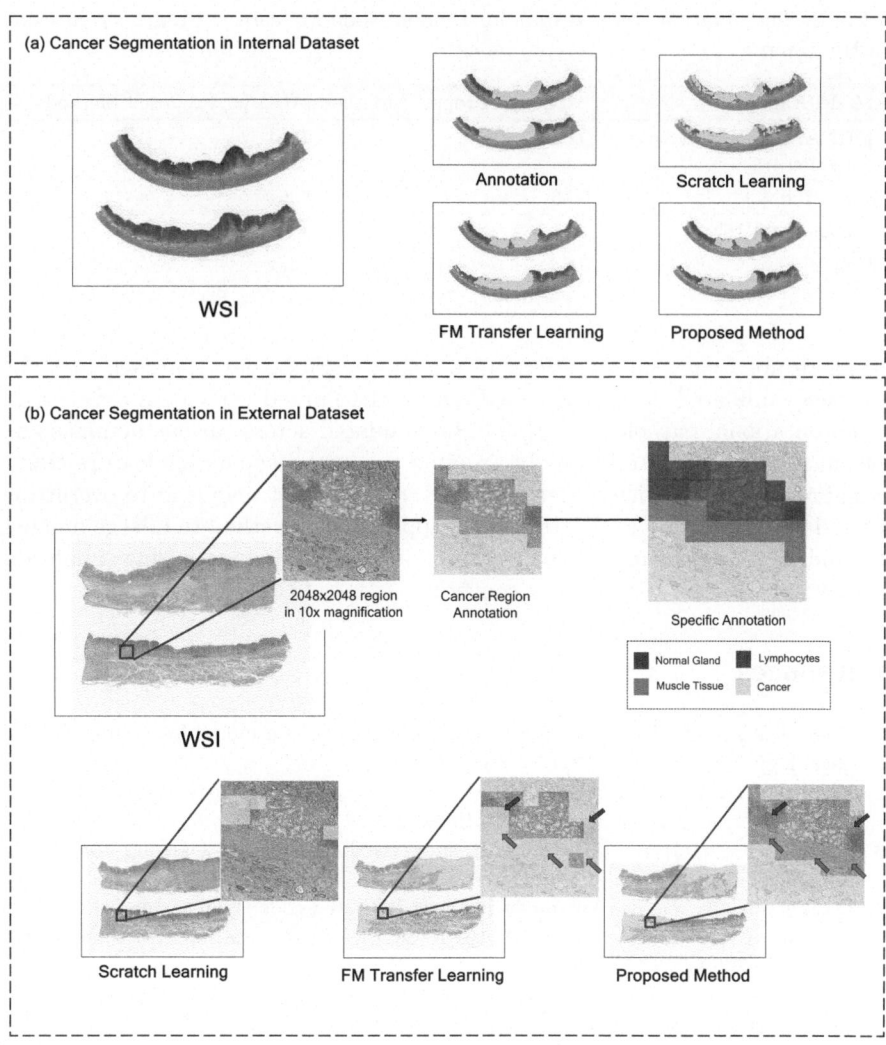

Fig. 3. Representative examples of cancer segmentation at the slide-level for scratch learning, the foundation model with transfer learning, and our proposed method. (a) Examples from the internal dataset. (b) Examples from the external dataset. The red arrow indicates the lymphocyte regions, and the green arrow indicates the muscle tissue regions. Our proposed method showed a reduction in the false positive areas around these patches. (Color figure online)

trained encoder (as shown in Fig. 1(b)). Additionally, our method can be flexibly applied not only to the foundation models used in this study but also to other pretrained networks suitable for various downstream tasks. This approach can be helpful when working with limited data, as is often the case in the medical domain.

Table 2. The ablation study results for Lymph Node Metastasis Prediction in Early Gastric Cancer.

Model Name	AUC w/o Proposed Method	AUC w/ Proposed Method
MONAI pathology tumor detection [15]	0.63	0.64
Ciga et al. [3]	0.56	0.61
CTransPath [16]	0.61	0.67
UNI [1]	0.63	0.71

There are some considerations in our study. First, our approach must be rigorously validated, as it was tested within the limited context of a few tasks in computational pathology. It should be evaluated across various domains and benchmark datasets. Additionally, training the autoencoders for feature transformation requires sufficient data; using a small dataset may lead to overfitting and reduced performance. Furthermore, applying our method to a large number of foundation models imposes more constraints, potentially requiring significant time and computing resources to converge.

References

1. Chen, R.J., et al.: Towards a general-purpose foundation model for computational pathology. Nature Med. **30**, 850–862 (2024)
2. Chen, T., Kornblith, S., Norouzi, M., Hinton, G.: A simple framework for contrastive learning of visual representations (2020)
3. Ciga, O., Martel, A., Xu, T.: Self supervised contrastive learning for digital histopathology (2020)
4. Dosovitskiy, A., et al.: An image is worth 16×16 words: transformers for image recognition at scale. arXiv:2010.11929 (2020). https://api.semanticscholar.org/CorpusID:225039882
5. He, K., Zhang, X., Ren, S., Sun, J.: Deep residual learning for image recognition. In: 2016 IEEE Conference on Computer Vision and Pattern Recognition (CVPR), pp. 770–778 (2015). https://api.semanticscholar.org/CorpusID:206594692
6. He, Y., et al.: Foundation model for advancing healthcare: challenges, opportunities, and future directions (2024)
7. kaiko.ai, Gatopoulos, I., Känzig, N., Moser, R., Otálora, S.: eva: evaluation framework for pathology foundation models. In: Medical Imaging with Deep Learning (2024). https://openreview.net/forum?id=FNBQOPj18N
8. Kang, M., Song, H., Park, S., Yoo, D., Pereira, S.: Benchmarking self-supervised learning on diverse pathology datasets. In: Proceedings of the IEEE/CVF Conference on Computer Vision and Pattern Recognition (CVPR), pp. 3344–3354 (June 2023)
9. Kramer, M.A.: Autossociative neural networks. Comput. Chem. Eng. **16**, 313–328 (1992). https://api.semanticscholar.org/CorpusID:62207837
10. Li, W., Peng, Y., Zhang, M., Ding, L., Hu, H., Shen, L.: Deep model fusion: a survey. arXiv:2309.15698 (2023). https://api.semanticscholar.org/CorpusID:262942062

11. Liu, Y., Zhang, W., Wang, J.: Adaptive multi-teacher multi-level knowledge distillation (2021)
12. Liu, Z., et al.: Swin transformer: hierarchical vision transformer using shifted windows. 2021 IEEE/CVF International Conference on Computer Vision (ICCV), pp. 9992–10002 (2021). https://api.semanticscholar.org/CorpusID:232352874
13. Mohammed, A., Kora, R.: A comprehensive review on ensemble deep learning: opportunities and challenges. J. King Saud Univ. Comput. Inform. Sci. **35** (2023). https://doi.org/10.1016/j.jksuci.2023.01.014
14. Oquab, M., et al.: DINOv2: learning robust visual features without supervision. arXiv:2304.07193 (2023). https://api.semanticscholar.org/CorpusID:258170077
15. Team, M.: Pathology tumor detection (2022). https://monai.io/model-zoo.html
16. Wang, X., et al.: Transformer-based unsupervised contrastive learning for histopathological image classification. Med. Image Anal. **81**, 102559 (2022). https://doi.org/10.1016/j.media.2022.102559
17. Zhang, S., Metaxas, D.: On the challenges and perspectives of foundation models for medical image analysis. Med. Image Anal. **91**, 102996 (2023). https://doi.org/10.1016/j.media.2023.102996
18. Zhang, Y., et al.: Data-centric foundation models in computational healthcare: a survey. arXiv preprint arXiv:2401.02458 (2024)
19. Zuchniak, K.: Multi-teacher knowledge distillation as an effective method for compressing ensembles of neural networks (2023)

OSATTA: One-Shot Automatic Test Time Augmentation for Domain Adaptation

Felix Küper and Sergi Pujades[✉]

University Grenoble Alpes, Inria, CNRS, Grenoble INP, LJK, Grenoble, France
sergi.pujades-rocamora@inria.fr

Abstract. Fundamental models (FM) are reshaping the research paradigm by providing ready-to-use solutions to many challenging tasks, such as image classification, registration, or segmentation. Yet, their performance on new dataset cohorts significantly drops, particularly due to domain gaps between the training (source) and testing (target) data. Recently, test-time augmentation strategies aim at finding target-to-source-mappings (t2sm), which improve the performance of the FM on the target dataset by leveraging the FM weights, thus assuming access to them. While this assumption holds for open research models, it does not for commercial ones (e.g., Chat-GPT). These are provided as black boxes; thus, the training data and the model weights are unavailable. In our work, we propose a new generic few-shot method that enables the computation of a target-to-source mapping by only using the black-box model's outputs. We start by defining a parametric family of functions for the t2sm. Using a simple loss function, we optimize the t2sm parameters based on a single labeled image volume. This effectively provides a mapping between the source domain and the target domain. In our experiments, we investigate how to improve the segmentation performance of a given FM (a UNet), and we outperform state-of-the-art accuracy in the 1-shot setting, with further improvement in a few-shot setting. Our approach is invariant to the model architecture as the FM is treated as a black box, which significantly increases our method's practical utility in real-world scenarios. The code is available for reproducibility purposes at https://osatta.gitlabpages.inria.fr/MedAGI.

1 Introduction

Fundamental models (FM) are reshaping the research paradigm by providing ready-to-use solutions to many challenging tasks, such as image classification, registration, or segmentation. The notion of a *fundamental model* is based on the hypothesis that the model is trained with such a high quantity and diversity of data, that it can generalize to any new dataset. In real-world scenarios, though, when dealing with small specific datasets, the performance of FM models severely deteriorates, mainly due to the domain gap between the training data used for the FM (source) and the new dataset (target). These domain gaps naturally arise from several sources. One source is the diversity in the acquisition devices; different vendor scanners or calibration parameters produce images

Z. Deng et al. (Eds.): MedAGI 2024, LNCS 15184, pp. 50–60, 2025.
https://doi.org/10.1007/978-3-031-73471-7_6

with different biases and noise patterns. A frequent domain gap also appears due to the differences in the observed patient population. These can be related to gender, age, ethnicity, anatomy, or pathologies. For these reasons, leveraging the potential of FM models on small specific datasets remains challenging to this day.

Interestingly, many different strategies have been studied in the literature to address this problem. They can be structured according to their stated hypothesis with regard to the nature of the FM and the target dataset. These hypotheses can be defined by answering the following questions: i) Is the training source data available? ii) What is the quantity of labeled and unlabeled target data samples? iii) Are the FM model weights available? Different combinations arise, which shape today's state-of-the-art. We start by describing approaches that try to make the FM as generic as possible by using data augmentation and then describe strategies to adapt them to target data.

Domain Generalization by Data Augmentation at Training Time. The goal here is to build the strongest possible model from the available data to best generalize to new unseen data. One widespread approach is data augmentation. However, choosing suitable image transformations is not trivial. Therefore, Cubuk et al. introduced AutoAugment [1] in which optimal augmentation strategies are automatically found. Since augmenting during training time is a bi-level optimization problem, their solution is highly compute-intensive. Later, Li et al. introduced 'Differentiable Automatic Data Augmentation" [2], in which they use the Gumbel-softmax parametrization trick to enable gradient descent over augmentation strategies. Follow-up work further optimized these methods, especially for the field of medical imaging [3–7].

Domain Adaptation. If the FM weights can be adapted using source and target domain data, we are in the field of Domain Adaptation (DA) [8]. If access to source data is not possible, we are in the setting of source-free domain adaptation (SFDA) [9] or source-free unsupervised domain adaptation (SFUDA) [10]. If sufficient labeled target data is available, Continual Learning strategies can be used [11]. These do not only allow to adapt a FM to a new domain, but also to new tasks. However, in a clinical context, the availability of labeled target data is limited, thus special strategies have been proposed. For example, Gaillochet et al. use uncertainty estimates during test-time adaptation to determine the best image to train on next [12], and Xu et al. introduced a gradual multi-stage technique to improve fine-tuning in low-resource scenarios [13]. If labeled target data is available, but data sharing is not possible, federated learning is a possibility. Li et al. introduced a technique that enables multi-site learning while preserving the privacy of the respective datasets [14]. For a more detailed look at this approach, we refer the reader to this survey [15]. Some methods assume access to unlabeled data from both target and source domain. Inspired by works such as CycleGan [16], their approach is to learn the t2sm directly from image pairs. With the recent advent of diffusion models, Gao et al. train a diffusion model to correct corruptions in images [17].

Test-Time Augmentation (TTA). Recently, TTA has been used to improve the robustness and accuracy of a FM without retraining/finetuning it. Not unlike data augmentation, the choice and weighting of different augmentation functions are non-trivial. Shanmugam et al. learned weighting functions for different augmentations to achieve new high scores on image-classification task [18]. For an extensive literature review on test-time adaptation strategies, we refer the reader to a recent survey [19]. Kimura et al. established a mathematical framework and showed the theoretical optimum of weighting augmentation functions [20]. In 2022, Tomar et al. introduced OptTTA [21], which showed that TTA can be used for domain adaptation. They exploit the fact that batch-normalization layers store statistics on the mean and variance of input data and feature layers. They use these weights to create an unsupervised loss function, optimizing augmentation strategies for the target dataset to achieve state-of-the-art accuracy. Their method is source-free and does not require access to any label for the target dataset. While they do not retrain the model, they assume access to the weights, which is key in the definition of their unsupervised loss function. In a follow-up work, You et al. introduced SaGTTA [22]. Instead of using the batch-norm layers information, they use saliency maps [23] to guide the optimization of the augmentation strategies. By minimizing the similarity between class saliency maps, they encourage confidence in predictions. Saliency maps also aid the interpretability of their results.

Both of these works [21,22] have inspired the proposed method. They assume no access to source data and optimize directly using unlabeled target data. However, they both assume access to the weights of the FM, which our approach, OSATTA, does not.

Annotating a full cohort of images is a very time-consuming task. However, the assumption that a few images (1-5) can be labeled, is a reasonable investment when dealing with a specific cohort. Our work also differs from OptTTA and SaGTTA in that regard; our approach, OSATTA, leverages a few labeled examples of the target domain in order to increase the performance of the FM in that specific cohort.

In summary, we propose OSATTA-for One Shot Automatic Test Time Augmentation, a one-shot approach that can adapt a black-box FM by finding a target-2-source mapping for the whole target dataset.

2   Method

Problem Statement. The input to our method is a fundamental model (FM) f performing a segmentation task in the image domain: $f(I) = S$, where, without loss of generality, $I \in \mathbb{R}^{W \times H}$ ($W, H \in \mathbb{N}$) is considered a 2D intensity image. The segmented image $S \in \mathbb{C}^{W \times H}$, has the same size as I with values spanning the set of possible $C \in \mathbb{N}$ class labels $\mathbb{C} = \{1, \ldots, C\}$. In addition to f we are provided with a small set of pairs of images and their labels, that we note $\mathcal{P} = \{(I_i, S_i)\}$, where $|\mathcal{P}| = N \in \mathbb{N}$ is small, typically $1 \leq N \leq 5$. Given a metric measuring the segmentation quality $D(f(I) = S, S^{GT})$, such as Dice score [24], our goal

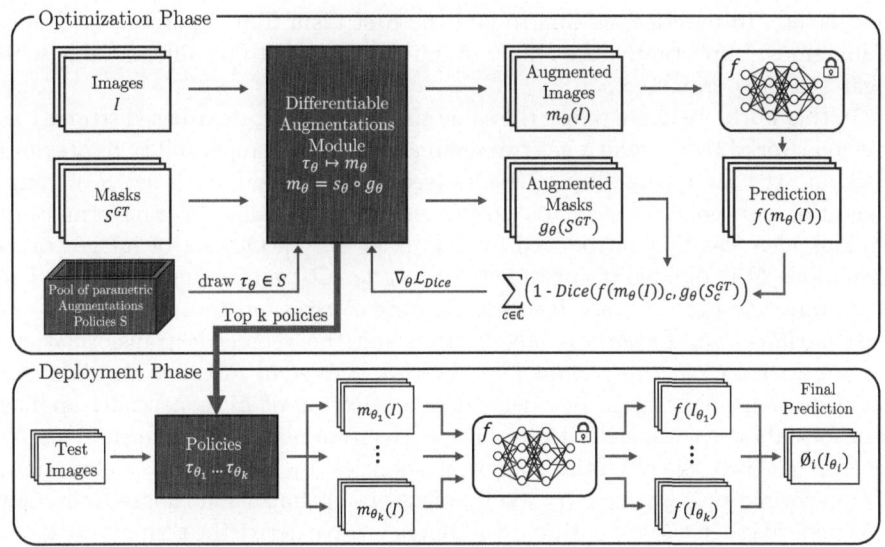

Fig. 1. Overview of our method. All possible policies from the pool get optimized using the available data. The top-k policies get picked to be used during the deployment phase, where we generate the final predictions.

is to find a target-to-source-mapping (t2sm) $m : \mathbb{R}^{W \times H} \to \mathbb{R}^{W \times H}$, composed of geometric transformation $(g : \mathbb{R}^{W \times H} \to \mathbb{R}^{W \times H})$ and style transformations $(s : \mathbb{R}^{W \times H} \to \mathbb{R}^{W \times H})$ so that $m = g \circ s$ and $D(f(m(I)), g(S^{GT})) \geq D(f(I), S^{GT})$. Where g is an invertible spatial transformation of the image. In practice, g can be effectively written as a product of multiple affine transformations, making it invertible. This invertibility is necessary to obtain the final segmentation mask on the original image frame, i.e., $g^{-1} \circ f \circ m(I) = S^m$. Note that in case g crops the image, then the inverse function has to assume a predetermined value for out-of-bounds values for the segmentation. We discuss this case in Sect. 4. The style function s is defined as a (potentially non-linear) change in the intensity values which does not affect the geometry of the image. Note that, unlike g, which can be applied to an image mask C, it is not possible to apply s to C.

The overview of OSATTA is presented in Fig. 1. In the rest of the section, we describe the set of elementary functions that we consider to build the s, g and thus m functions (Sect. 2.1), we explain how we find/optimize the elementary function combinations (Sect. 2.2), and how these policies are ensembled to obtain our results at test-time (Sect. 2.3).

2.1 T2sm Augmentation Space

The t2sm could theoretically be any invertible function that projects the image back into the original image space as in $m : \mathbb{R}^{W \times H} \mapsto \mathbb{R}^{W \times H}$. This includes very complex non-linear mappings, such as neural networks, CNNs, diffusion models, etc. In practice, we aim to find the best t2sm from a parametric family of functions. Importantly, the family of functions needs to be differentiable with regard to the parameters.

In this work, we chose to use the same simple augmentation-based strategy as previous work [21,22], which has two main advantages: comparability to previous work and the interpretability of results (see sect. 3). We build a set S of "augmentation sub-policies" τ, made up of elementary parametric transformations $\mathcal{O}$. This idea was first introduced by Cubuk et al. [1]. Our set of image transformations $\mathcal{O}$ is divided into two sub-sets: style ($\mathcal{O}_s$) and geometric ($\mathcal{O}_g$). The style transformations $\mathcal{O}_s$ are *Identity, Gamma Correction, Gaussian Blur, Contrast modification, Brightness modification* and the geometric transformations $\mathcal{O}_g$ are *Resize Crop, Horizontal Flip, Vertical Flip, and Random Rotation*. An augmentation policy τ_θ is then defined as a selection of N sequentially applied transformations parametrized by θ. For a given number of transformations N, we then generate the potential pool of sub-policies, by building all combinations of transformations, giving us $\binom{9}{N}$ sub-policies to optimize and choose from. Our final policy is then a set of optimized sub-policies. We derive the t2sm $m_\theta = g_\theta \circ s_\theta$ from the sub-policy τ_θ with the parameters θ by sequentially applying the parametric transformations, therefore $\tau_\theta \mapsto m_\theta$

2.2 Supervised Optimization of Augmentations

Differentiable Augmentation Functions. As the image transformations themselves are not directly differentiable, we have to use a "re-parametrization trick": each sub-policy τ is optimized by minimizing the expected loss $\mathcal{L}_\tau$, which is the expectation of the loss over random augmentations of the data. By expressing the augmentation magnitudes in terms of learnable distribution parameters μ^τ and σ^τ, we enable the numeric estimation of gradients for these parameters. For a detailed description, we refer the reader to Sect. 2.2.2 of OptTTA [21].

Training Phase. Our training regime is a modified version of the one used in OptTTA [21]. We next provide an overview, pointing out the key differences. During training, given a number of maximum augmentations, we generate our pool of sub-policies as described in sec. 2.1. For a given number of iterations, we then optimize the parameters θ of each sub-policy in the following way. First, we draw mini-batches of image slices across our entire training set. This is a key difference compared to OptTTA and SaGTTA, which optimize one image volume at a time. By optimizing across multiple images, we aim to generalize better to new unseen images. Then, we augment the mini-batch using the current sub-policy (with optimized parameters θ). Importantly, for all geometric transformations, we use the same seed to augment the matching segmentation masks. As discussed earlier, we do not apply any style transformations to the segmentation masks, as we assume that any intensity change should not affect the segmentation labels. We then generate predictions for our current augmented mini-batch and calculate the loss using the augmented segmentation masks. We

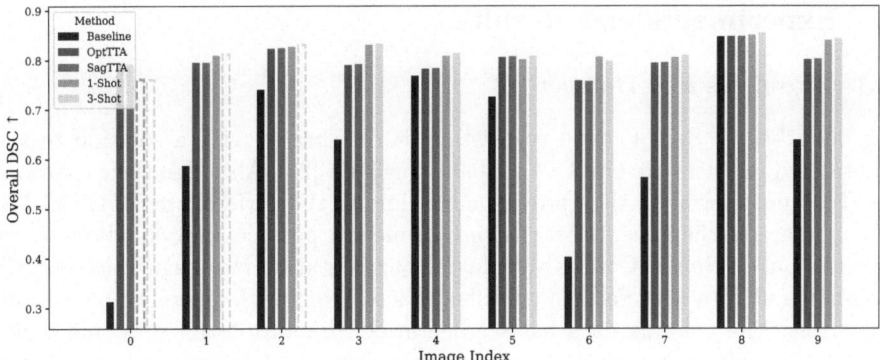

Fig. 2. Individual Dice scores for each image, compared across methods, dotted lines for images used during few-shot training (score provided for completeness)

use the Dice loss [25]

$$\mathcal{L}(I, S^{GT}; \theta) = \sum_{c \in \mathbb{C}} (1 - Dice(f(m_\theta(I))_c, g_\theta(S_c^{GT}))) \tag{1}$$

to optimize the parameters θ of the current sub-policy via gradient descent. We train our optimization policies using AdamW optimizer for 3k iterations at a learning rate of 10^{-3}. We implemented all experiments using PyTorch and ran them on an NVIDIA RTX 2080 Ti.

2.3 Deployment Phase

After optimizing the entire pool of possible augmentation sub-policies, we select the top k sub-policies to use during deployment. The selection criterion here is Dice loss of the training image. To reduce overfitting, we considered using an additional labeled image volume to compute a validation loss as a selection criterion. However, in practice, including this image in the training set and using the training loss as a criterion proved to generalize better to the test set. In contrast to previous (unsupervised) works, we do not fine-tune the top policies on every test image during deployment. In our evaluation, we discuss the performance and accuracy implications of this (see sect. 3). Besides that, the ensembling of the final prediction is done in the same fashion as SaGTTA [22]. We draw M random samples of each sub-policy, average the M results, and obtain an estimate of the expected value for the given sub-policy. We then average over all k sub-policies, obtaining our final prediction for the total policy made up of our top-k optimized sub-policies.

3   Experiments and Results

3.1   Baselines and Dataset

As described in Sect. 1, the two existing works that provide a solution to the considered problem are OptTTA [21] and SagTTA [22]. Although their solutions are fully unsupervised, they provide a baseline to the performance of OSATTA. To compare to them we follow their experimental protocol. For the dataset, we use the public Spinal Cord Gray Matter Segmentation challenge Dataset [26], consisting of data collected at four different sites, each with different vendors and protocols. The training data has segmentation annotations for gray and white matter, and all images were re-sampled to a common 1mm isotropic resolution using bi-cubic interpolation. For the experimental setting, we focus on adapting the data from site #3 to a model with fixed weights trained on data from site #1. As described in SaGTTA [22], these two sites happen to show the most significant domain gap of the dataset. The learned model, considered as the FM, is a 2D U-net trained with weighted cross entropy loss with the RMSprop optimizer (learning rate 10^{-5}; 250K iterations). The segmentation performance on left-out data is measured using the Dice score (DSC) [24].

Table 1. Comparison of DSC score for different methods on target test data. FM trained on site 1, and t2sm learned with target site 3. Bold indicates the best value, and underline the second best.

Method	FM no adapt.	OptTTA	SaGTTA	SaGTTA no exploit	1-Shot (ours)	3-Shot (ours)
DSC Mean	.5709	.7943	.8016	.7806	<u>.8197</u>	**.8229**
DSC Std	.0200	.0416	.0275	.0601	<u>.0249</u>	**.0209**

Values for the baselines (Baseline, OptTTA, SaGTTA) were obtained from [22], as these results were averaged over 20 runs. Due to time constraints, we were unable to replicate their extensive number of runs. Nonetheless, our reproduced values are consistent with the literature, falling within a similar range.

3.2   Evaluation

In Table 1, we report the computed mean DSC values between the GT segmentation masks and the predicted ones. Our approach obtains the highest values, with a significant improvement with respect to the FM (no adapt.) in the one-shot scenario. Unlike the baselines, our approach leverages the availability of more labeled training target data, as visible in the 3-shot approach, slightly increasing the performance over the 1-shot. In Fig. 2, we present the performance on the individual target test data, where one can observe how our method bridges the domain gap for the other target images that have not been seen. Interestingly, we outperform SagTTA, even though it was further fine-tuned individually on each image. It is worth noting that SagTTA [22] obtains, via a thorough grid

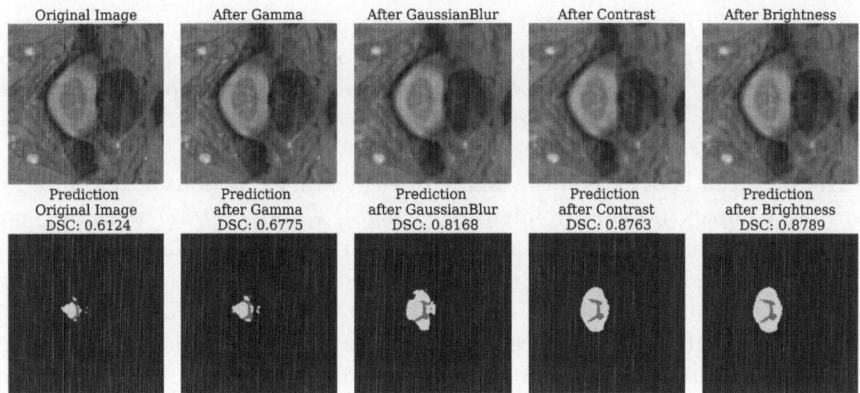

Fig. 3. Visualization of the learned top policy: Top row: Image after augmentation, Bottom row: Prediction of the augmented image above

search, a theoretical upper bound of .8036 for the DSC performance. Both our approaches outperform this upper bound with few examples (.8229).

One interesting feature of our approach, inherited from the selection of the set of policies [21], is that their effects can be visually observed. In the top of Fig. 2, we present, from left to right, the original image and the different transformations applied until the final image. On the bottom, we observe the applied FM to the corresponding image. It is interesting to note how each transformation further improves the prediction of the FM.

Speed. We further analyze the speed of the application of the approaches. During training, we obtain equal iteration speed to OptTTA [21] with ~ 4 it/s. SaGTTA [22] is one order of magnitude slower with 0.5 it/s. However, as we do not change the ensembling process of the trained augmentation policies for the final prediction, our method is orders of magnitude faster at test time. While the accuracy of previous methods relies on fine-tuning policies for each image, we can skip this step and apply the learned augmentation directly (Fig. 3).

4   Conclusion

In this work, we propose OSATTA: One Shot Automatic Test Time Adaptation. OSATTA learns a target-2-souce-mapping (t2sm) to improve the segmentation accuracy of a Fundamental Model (FM) on a new target domain, without requiring access to the weights of the FM. By leveraging one labeled sample of the target domain, OSATTA improves the performance of the FM model by over .2 points DICE and also outperforms existing unsupervised state-of-the-art approaches. OSATTA learns a t2sm from a set of simple parametrized augmentation functions. The definition of this set allows to obtain an interpretability of the learned function, and thus visualize the domain gap between the source and target domains.

While OSATTA is presented with a segmentation tasks, it is interesting to note that OSATTA can be easily extended to other tasks, such as registration or classification. To do so, one only needs to replace the output quality measure function, so that it is aligned with the considered task. We leave these experimentations for future work.

Another lead is to extend the family of elementary functions to optimize for. Learning complex non-linear intensity changes, adding 2D/3D warpings, and parametrizing the current non-learnable geometric transformations are just some of the possibilities that could be explored. Moreover, one could try to learn a full style transfer model, such as a U-Net. Although this seems a difficult task in the low data regime (few-shot), one could potentially go beyond pre-defined functions and learn other patterns between the source and target domain. OSATTA opens a new perspective on how to improve the fundamental models performance in unseen target domains.

Acknowledgements. Felix Küper and Sergi Pujades' work was funded by the ANR PRC INORA project.

Disclosure of Interests. All authors have no competing interests to declare that are relevant to the content of this article.

References

1. Cubuk, E.D., Zoph, B., Mane, D., Vasudevan, V., Le, Q.V.: AutoAugment: learning augmentation strategies from data. In: 2019 IEEE/CVF Conference on Computer Vision and Pattern Recognition (CVPR), pp. 113–123. IEEE, Long Beach, CA, USA (2019). https://doi.org/10.1109/CVPR.2019.00020
2. Li, Y., Hu, G., Wang, Y., Hospedales, T., Robertson, N.M., Yang, Y.: Differentiable automatic data augmentation. In: Vedaldi, A., Bischof, H., Brox, T., Frahm, J.M. (eds.) Computer Vision - ECCV 2020, pp. 580–595. Springer International Publishing, Cham (2020). https://doi.org/10.1007/978-3-030-58542-635
3. He, W., Liu, M., Tang, Y., Liu, Q., Wang, Y.: Differentiable automatic data augmentation by proximal update for medical image segmentation. IEEE/CAA J. Automatica Sinica **9**(7), 1315–1318 (2022). https://doi.org/10.1109/JAS.2022.105701
4. Liu, A., Huang, Z., Huang, Z., Wang, N.: Direct differentiable augmentation search. In: 2021 IEEE/CVF International Conference on Computer Vision (ICCV), pp. 12199–12208. IEEE, Montreal, QC, Canada (2021). https://doi.org/10.1109/ICCV48922.2021.01200
5. Xu, J., Li, M., Zhu, Z.: Automatic data augmentation for 3D medical image segmentation (2020). https://doi.org/10.48550/arXiv.2010.11695, arXiv:2010.11695
6. Liu, Z., Lv, Q., Li, Y., Yang, Z., Shen, L.: MedAugment: universal automatic data augmentation plug-in for medical image analysis (2023). https://doi.org/10.48550/arXiv.2306.17466, http://arxiv.org/abs/2306.17466
7. Luo, Y., Wang, Y., Zhang, Z., Liu, M., Tang, Y.: IOADA: An optimal automated augmentation algorithm for medical image segmentation. In: 2023 China Automation Congress (CAC), pp. 3900–3905 (2023). https://doi.org/10.1109/CAC59555.2023.10450689, iSSN: 2688-0938

8. Guan, H., Liu, M.: Domain adaptation for medical image analysis: a survey. IEEE Trans. Biomed. Eng. **69**(3), 1173–1185 (2022). https://doi.org/10.1109/TBME.2021.3117407, https://ieeexplore.ieee.org/document/9557808/

9. Yu, Z., Li, J., Du, Z., Zhu, L., Shen, H.T.: A comprehensive survey on source-free domain adaptation (2023). https://doi.org/10.48550/arXiv.2302.11803, arXiv:2302.11803 [cs]

10. Fang, Y., Yap, P.T., Lin, W., Zhu, H., Liu, M.: Source-free unsupervised domain adaptation: a survey (2023). https://doi.org/10.48550/arXiv.2301.00265, arXiv:2301.00265

11. Wang, L., Zhang, X., Su, H., Zhu, J.: A comprehensive survey of continual learning: theory, method and application (2024). https://doi.org/10.48550/arXiv.2302.00487, http://arxiv.org/abs/2302.00487, arXiv:2302.00487

12. Gaillochet, M., Desrosiers, C., Lombaert, H.: TAAL: test-time augmentation for active learning in medical image segmentation. In: Nguyen, H.V., Huang, S.X., Xue, Y. (eds.) Data Augmentation, Labelling, and Imperfections, pp. 43–53. Springer, Cham (2022). https://doi.org/10.1007/978-3-031-17027-0_5

13. Xu, H., Ebner, S., Yarmohammadi, M., White, A.S., Van Durme, B., Murray, K.: Gradual fine-tuning for low-resource domain adaptation. arXiv:2103.02205 (2021)

14. Li, X., Gu, Y., Dvornek, N., Staib, L., Ventola, P., Duncan, J.S.: Multi-site fMRI analysis using privacy-preserving federated learning and domain adaptation: ABIDE Results (2020). http://arxiv.org/abs/2001.05647

15. Zhang, C., Xie, Y., Bai, H., Yu, B., Li, W., Gao, Y.: A survey on federated learning. Knowl. Based Syst. **216**, 106775 (2021). https://doi.org/10.1016/j.knosys.2021.106775, https://www.sciencedirect.com/science/article/pii/S0950705121000381

16. Zhu, J.Y., Park, T., Isola, P., Efros, A.A.: Unpaired image-to-image translation using cycle-consistent adversarial networks. arXiv:1703.10593 (2020)

17. Gao, J., Zhang, J., Liu, X., Darrell, T., Shelhamer, E., Wang, D.: Back to the source: diffusion-driven adaptation to test-time corruption. In: 2023 IEEE/CVF Conference on Computer Vision and Pattern Recognition (CVPR), pp. 11786–11796. IEEE, Vancouver, BC, Canada (2023). https://doi.org/10.1109/CVPR52729.2023.01134

18. Shanmugam, D., Blalock, D., Balakrishnan, G., Guttag, J.: Better aggregation in test-time augmentation. arXiv:2011.11156 (2021)

19. Liang, J., He, R., Tan, T.: A comprehensive survey on test-time adaptation under distribution shifts (2023). https://doi.org/10.48550/arXiv.2303.15361, http://arxiv.org/abs/2303.15361, arXiv:2303.15361

20. Kimura, M.: Understanding test-time augmentation (2024). https://doi.org/10.48550/arXiv.2402.06892, http://arxiv.org/abs/2402.06892

21. Tomar, D., Vray, G., Thiran, J.P., Bozorgtabar, B.: OptTTA: learnable test-time augmentation for source-free medical image segmentation under domain shift. In: Proceedings of Machine Learning Research, Volume 172: International Conference on Medical Imaging with Deep Learning, 6-8 July 2022, Zurich, Switzerland (2022)

22. You, S., Tomar, D., Bozorgtabar, B., Reyes, M.: SaGTTA: saliency guided test time augmentation for medical image segmentation across vendor domain shift. In: 2023 IEEE 20th International Symposium on Biomedical Imaging (ISBI), pp. 1–4 (2023). https://doi.org/10.1109/ISBI53787.2023.10230764, iSSN: 1945-8452

23. Sundararajan, M., Taly, A., Yan, Q.: Axiomatic attribution for deep networks (2017). https://doi.org/10.48550/arXiv.1703.01365, http://arxiv.org/abs/1703.01365, arXiv:1703.01365

24. Dice, L.R.: Measures of the amount of ecologic association between species. Ecology **26**(3), 297–302 (1945). https://doi.org/10.2307/1932409, https://onlinelibrary. wiley.com/doi/abs/10.2307/1932409

25. Sudre, C.H., Li, W., Vercauteren, T., Ourselin, S., Cardoso, M.J.: Generalised dice overlap as a deep learning loss function for highly unbalanced segmentations, vol. 10553, pp. 240–248 (2017). https://doi.org/10.1007/978-3-319-67558-9_28, http:// arxiv.org/abs/1707.03237, arXiv:1707.03237

26. Prados, F., et al.: Spinal cord grey matter segmentation challenge. Neuroimage **152**, 312–329 (2017). https://doi.org/10.1016/j.neuroimage.2017.03.010, https:// www.sciencedirect.com/science/article/pii/S1053811917302185

Automating MedSAM by Learning Prompts with Weak Few-Shot Supervision

Mélanie Gaillochet[1,2,3]([✉]), Christian Desrosiers[1], and Hervé Lombaert[1,2,3]

[1] ÉTS Montréal, Montréal, Canada
[2] Polytechnique Montréal, Montréal, Canada
[3] Mila - Quebec AI Institute, Université de Montréal, Montréal, Canada
melanie.gaillochet.1@ens.etsmtl.ca

Abstract. Foundation models such as the recently introduced Segment Anything Model (SAM) have achieved remarkable results in image segmentation tasks. However, these models typically require user interaction through handcrafted prompts such as bounding boxes, which limits their deployment to downstream tasks. Adapting these models to a specific task with fully labeled data also demands expensive prior user interaction to obtain ground-truth annotations. This work proposes to replace conditioning on input prompts with a lightweight module that directly learns a prompt embedding from the image embedding, both of which are subsequently used by the foundation model to output a segmentation mask. Our foundation models with learnable prompts can automatically segment any specific region by 1) modifying the input through a prompt embedding predicted by a simple module, and 2) using weak labels (tight bounding boxes) and few-shot supervision (10 samples). Our approach is validated on MedSAM, a version of SAM fine-tuned for medical images, with results on three medical datasets in MR and ultrasound imaging. Our code is available on https://github.com/Minimel/MedSAMWeakFewShotPrompt Automation.

Keywords: Large Vision Models · Segmentation · Medical · Prompt

1 Introduction

Annotation is a well-known labour-intensive and time-consuming task in medical imaging. Supervised segmentation models trained to identify specific regions of interest do not generalize well to new domains or classes and require more data and retraining when considering a new task. This increases the cost of developing segmentation models to solve multiple tasks. The need for universal models that can be applied to various tasks after training has hence been growing in medical image analysis. The introduction of foundation models for image segmentation such as the recent Segment Anything Model (SAM) [10], as well as its

Z. Deng et al. (Eds.): MedAGI 2024, LNCS 15184, pp. 61–70, 2025.
https://doi.org/10.1007/978-3-031-73471-7_7

versions adapted for medical imaging [21], notably MedSAM [12], have appeared as a game-changer in the field of computer vision and medical image analysis. These models have shown remarkable performance on a variety of segmentation tasks. However, they remain promptable models that require user interaction to obtain the segmentation mask of a target object. Furthermore, their zero-shot performance depends on the quality of the user prompt. This reliance on user interaction hinders their integration into automatic pipelines and limits their usability at a large scale.

Recent attempts have been made to automate the prompt generation of SAM [17,20,22]. However, these methods typically require samples with ground-truth segmentation masks, which are costly to obtain in the medical domain.

This paper proposes a lightweight add-on prompt module which learns to generate prompt embeddings directly from SAM's image embedding. Our end-to-end approach enables SAM models to specialize on the segmentation of a specific region and only requires few weakly-annotated samples. This reduces the interaction cost of developing specialized segmentation models. Our validation shows that, given only few training samples weakly annotated with tight boxes, promptable foundation model can effectively generate segmentation masks of target regions without requiring manual prompt inputs.

Foundation Models for Medical Image Segmentation. Vision foundation models have achieved tremendous success in computer vision tasks thanks to large-scale pre-training. In particular, the Segment Anything Model (SAM) [10], based on vision transformers [5] and trained on 1B masks and 11M images, was recently introduced as a prompt-driven foundation model for segmentation. Trained on natural images, SAM obtains uneven performances on medical data [7,13,18], inducing its adaptation to the medical domain [4,12,17]. In particular, MedSAM [12], a foundation model for universal medical image segmentation was trained on 1.5 million image-mask pairs over 10 imaging modalities. These models provide impressive zero-shot performance, but remain promptable models that require user interaction at inference.

Prompt Automation for SAM. Motivated by its performance in Natural Language Processing [2], prompt-tuning has successfully been applied to large vision models [8]. Hence, methods that have focused on specializing SAM, a promptable model, have naturally explored prompt generation. Given few fully labeled samples, the self-prompting unit of [20] automatically generates a real point and bounding box from SAM's image embedding. AutoSAM replaces SAM's prompt encoder with a Harmonic Dense Net to adapt segmentation to medical images [17]. A recent training-free approach, PerSAM [22], encodes positive-negative location priors as prompt tokens to produces automatic segmentations of a specific object from a single reference image and mask. As opposed to our approach, all of these methods require samples with full segmentation masks.

Segmentation with Bounding Box Annotations. Bounding boxes have emerged as an alternative to onerous annotation masks. Most methods use bounding boxes as an initial pseudo-label of the target region. A classic iterative graph-cut-based algorithm, GrabCut [16], separates the foreground from its background given a bounding box. DeepCut [14] extends GrabCut to neural networks using existing heuristics. More recently, the bounding box tightness prior was adapted to deep learning-based models by imposing a set of constraints on the predictions [9], and was combined with multiple instance learning and smooth maximum approximation [19].

Our Contribution. This work aims to efficiently automate MedSAM, a variant of SAM for the medical domain, to segment any target region through the use of few, weakly-labeled samples. Our approach introduces an innovative improvement by substituting the original prompt encoder, which requires user input, with an enhanced lightweight adaptable prompt-learning module that:

1. Automatically **generates a prompt embedding** from the input image
2. Trains with only **weak labels** (tight bounding boxes) and **few-shot** learning
3. Is easily added on top of MedSAM (no fine-tuning)

The next sections present our proposed prompt module for MedSAM and demonstrate its usefulness on various medical image segmentation tasks.

2 Methodology

2.1 Preliminaries: MedSAM Architecture

Our approach builds upon on MedSAM [12], a variant of SAM [10] fine-tuned on medical data. The model has three main components: a large image encoder E_{img}, a prompt encoder E_{pr} and a lightweight mask decoder D_{mask}.

While the image encoder computes an embedding of the input image x, the prompt encoder outputs two sparse and dense embeddings from the provided set of prompts $[pr]$, respectively points or bounding boxes (BB), and a mask. The network produces a probability map f_θ by taking x and a prompt embedding $Z_{pr} = E_{pr}([pr])$:

$$f_\theta = \sigma\big(D_{mask}(E_{img}(x), Z_{pr})\big),$$

where σ is the sigmoid function.

We present an end-to-end approach to remove the typical dependence on user-defined prompts $[pr]$, without modifying the pretrained MedSAM network.

2.2 Lightweight Prompt Module

Our approach consists of a prompt module trained to compute directly Z_{pr} from the image embedding provided by MedSAM (see Fig. 1b). The module outputs two embeddings of the same shape as those generated by MedSAM (Fig. 1a). Originally, the dense prompt embedding has a spatial correspondence with the image and can be considered as a low-quality segmentation map, while the sparse embeddings are spatial encodings of coordinates. Therefore, our prompt module generates a dense embedding through a convolutional layer and a sparse embedding through a fully connected (FC) layer.

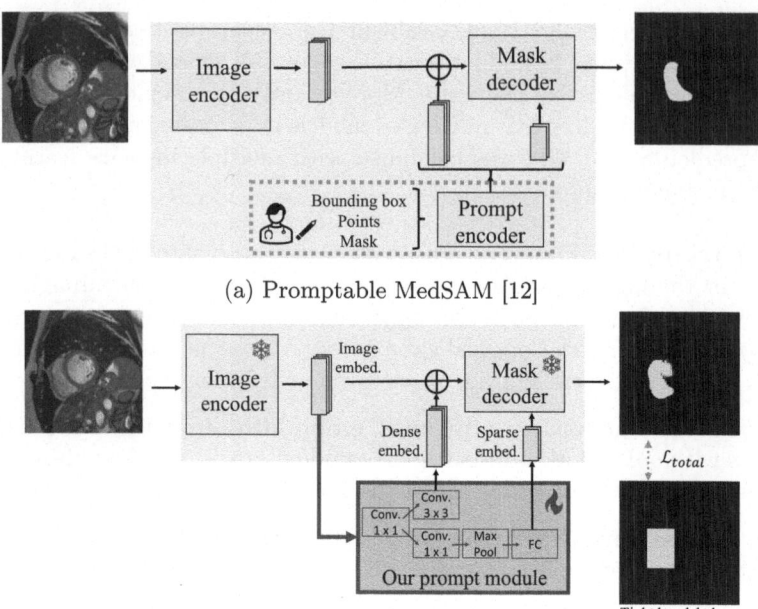

(a) Promptable MedSAM [12]

(b) Automatic MedSAM (ours)

Fig. 1. Comparison between (a) MedSAM and (b) our automation of MedSAM via a learnt prompt module. Our prompt module replaces MedSAM's prompt encoder and learns to generate a relevant prompt embedding from the image embedding. Training employs losses that utilize only tight box labels.

2.3   Learning with Tight Box Annotations

Denote as $X : \Omega \subset \mathbb{R}^{3 \times H \times W} \to \mathbb{R}$ a 3-channel input image of height H and width W, where Ω is the spatial domain corresponding to each channel of the image. Moreover, let $Y \in \{0, 1\}^{\Omega}$ be the ground-truth binary segmentation mask of X. Suppose we only have access to a tight bounding box $\tilde{Y}$ of the target. Ω_I and Ω_O define the regions respectively inside and outside the bounding box such that $\Omega_I + \Omega_O = \Omega$. This leads to a constrained optimization problem [9] from the bounding box annotations $\tilde{Y}$.

Emptiness of Ω_O. Since the region in Ω_I defined by the bounding box must contain the target object, Ω_O must contain only foreground. Hence, we can apply a Cross-entropy loss for all pixels $p \in \Omega_O$:

$$\mathcal{L}_{empty} = - \sum_{p \in \Omega_O} \log(1 - f_{\theta}(p)). \tag{1}$$

Tight Box Constraint in Ω_I. The tightness of the bounding box indicates that at least one foreground pixel must cross every horizontal and vertical line of weak label $\tilde{Y}$. As in [9], we soften this condition by considering segments of width w instead of individual lines and ensure differentiability by considering output probabilities instead of the prediction mask. The condition formalizes as:

$$\sum_{p \in s_l} f_\theta(p) \geq w, \quad \forall s_l \in S_L, \tag{2}$$

where S_L is the set of all vertical and horizontal segments of width w that make the bounding box $\tilde{Y}$. We convert the inequality constraints of (2) to a loss using a penalty function ψ_t, and obtain:

$$\mathcal{L}_{tightbox} = \sum_{s_l \in S_L} \psi_t \left(w - \sum_{p \in s_l} f_\theta(p) \right). \tag{3}$$

The penalty function can be modeled as a simple scaled ReLU function, i.e. $\psi_t(x) = t \cdot \max(0, x)$. In this work, we instead resort to a pseudo log-barrier function, which provides a more stable optimization under multiple competing constraints. As $t \to \infty$, function $\psi_t(x)$ behaves as a hard barrier where $\psi_t(x) = \infty$ if $x > 0$, else $\psi_t(x) = 0$. In our method, we found that using a fixed value of $t = 5$ worked best.

Foreground Size Constraint. The bounding box $\tilde{Y}$ also sets a limit on the target size of the prediction mask. Again, we consider output probabilities rather than individual predictions to ensure differentiability. By applying priors on the fraction $\epsilon \in [0, 1]$ of pixels from Ω_I that belong to the background, we get:

$$\epsilon_1 |\Omega_I| \leq \sum_{p \in \Omega} f_\theta(p) \leq \epsilon_2 |\Omega_I|. \tag{4}$$

As before, we employ $\psi_t(x)$ to convert these inequality constraints into the following loss:

$$\mathcal{L}_{size} = \psi_t \left(\epsilon_1 |\Omega_I| - \sum_{p \in \Omega} f_\theta(p) \right) + \psi_t \left(\sum_{p \in \Omega} f_\theta(p) - \epsilon_2 |\Omega_I| \right). \tag{5}$$

Given (1), (3) and (5), and weights λ_1 and λ_2, the final loss becomes:

$$\mathcal{L}_{total} = \mathcal{L}_{empty} + \lambda_1 \mathcal{L}_{tightbox} + \lambda_2 \mathcal{L}_{size}. \tag{6}$$

3 Results

3.1 Datasets

Our experiments validate our method on three public datasets: the Head Circumference dataset[1] (HC18) [6], the Cardiac Acquisitions for Multi-structure Ultrasound Segmentation[2] and the Automated Cardiac Diagnosis Challenge[3]

---

[1] https://hc18.grand-challenge.org/
[2] https://www.creatis.insa-lyon.fr/Challenge/camus/ (CAMUS) [11]
[3] https://humanheart-project.creatis.insa-lyon.fr/database/

(ACDC) [1]. For both cardiac datasets, the end diastole images are used. For HC18, we filter out samples with ground-truth masks that could not be automatically generated by OpenCV from the circumference annotations, and split the ultrasound dataset into 507 training, 77 validation and 148 test images. For CAMUS, we focus on the left ventricle (LV) and left atrium (LA) segmentation and use 50 images for validation, 100 images for testing and the remaining 350 images for training. For ACDC, we focus on the right ventricule (RV) and LV segmentation and use 10 patients for validation (78 images), 50 for testing (470 images)and the remaining 90 patients (765 images) for training. Each sample has a minimum foreground size for all experiments.

Following [12], our preprocessing includes clipping the intensity values of each 2D image (HC18, CAMUS) or each 3D volume (ACDC) between the 0.5^{th} and 99.5^{th} percentiles and rescaling them to the range $[0, 255]$. We also partition each 3D volume of ACDC into 2D image which we resample to a fixed $1mm \times 1mm$ resolution. We center crop and pad each sample to size 640×640 (HC), 512×512 (CAMUS) or 256×256 (ACDC). To meet MedSAM's requirements, we resize all images to a fixed $3 \times 1024 \times 1024$ size before inputting them in the model.

3.2   Implementation Details

Model. Our backbone model is MedSAM based on ViT-B, the smallest version of SAM. The backbone remains frozen during training. We keep our prompt module lightweight by using few layers. A 1×1 convolution first reduces the number of channels. Then, the dense embedding is obtained through a 3×3 convolution, while the sparse embedding is obtained through a 1×1 convolution followed by max pooling and a fully connected layer. All convolutional layers are followed by ReLU activation. Our prompt module has thus 2.4 M trainable parameters.

Loss Parameters. We train our prompt module using $\mathcal{L}_{total}$, with $\lambda_1 = 0.0001$, $\lambda_2 = 0.01$. For our tight box constraint, we follow [9] and use segments of $w = 5$. We hypothesize that the foreground region is at least half the size its tight bounding box and set $[\epsilon_1, \epsilon_2] = [0.5, 0.9]$. Comparative training with full segmentation masks uses a Binary Cross-entropy Dice loss, each term having the same weight.

Training. We use a batch size of 4 and a learning rate (LR) of 0.001 with a multi-step scheduler decreasing LR by 0.1 after half the epochs and a weight decay of 0.0001. To minimize computational complexity, we do not use data augmentation. This allows us to discard MedSAM's image and prompt encoders after saving the image embeddings during an initial iteration, reducing the number of total parameters from 96.1 M to only 6.5 M (2.4 M trainable). In the 10-shot setting, we repeat the experiments 9 times, with 3 initialization seeds and 3 training subsets selected uniformly at random. The results are averaged over these experiments. All experiments are implemented in Python 3.8.10 with Pytorch on NVIDIA RTX-A6000 GPUs.

Baselines. For each class, we compare our method with two specialized single-task models: a standard UNet [15] and a TransUNet [3]. We also validate our method against PerSAM [22] which automates SAM with 1-shot supervision,

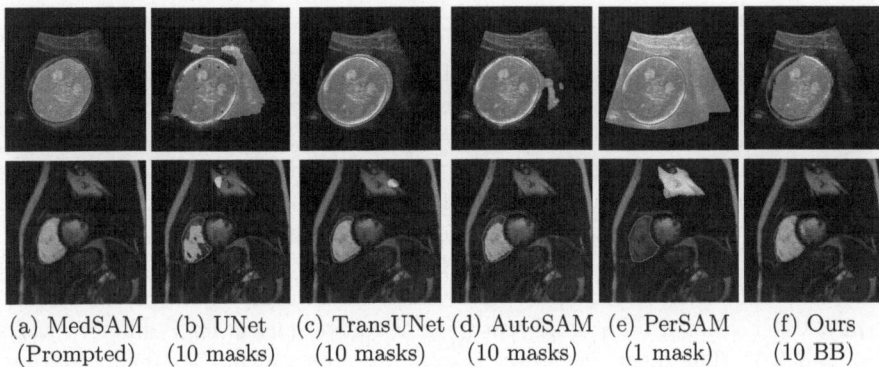

(a) MedSAM	(b) UNet	(c) TransUNet	(d) AutoSAM	(e) PerSAM	(f) Ours
(Prompted)	(10 masks)	(10 masks)	(10 masks)	(1 mask)	(10 BB)

Fig. 2. Predicted segmentations on test samples of HC18 (row 1) and the right ventricle in ACDC (row 2). From left to right, (a) MedSAM prompted with a tight box, (b–d) UNet, TransUNet and AutoSAM, trained with ground-truth masks, (e) PerSAM using one reference image with its ground-truth mask, and (f) our method trained on tight bounding boxes. All automatic methods are given for the 10-shot setting, except PerSAM, a 1-shot approach. Ground-truth annotation is drawn in red, with predicted segmentation mask overlayed in yellow. (Color figure online)

Table 1. Model performance on test sets in terms of mean (std) 2D Dice similarity score (↑). Best results in few-shot settings are highlighted in bold

Type	Method	# Samples	Mask labels	BB labels	HC	CAMUS LV	CAMUS LA	ACDC RV	ACDC LV
Promptable	MedSAM [12] (w/ tight box)	–	–	–	95.19	94.50	89.23	93.78	95.45
Automatic (fully trained)	UNet [15]	All	✓		$86.53_{\pm0.55}$	$89.93_{\pm0.01}$	$74.77_{\pm0.78}$	$89.55_{\pm0.23}$	$94.83_{\pm0.13}$
		10	✓		$61.79_{\pm3.10}$	$75.09_{\pm3.69}$	$46.29_{\pm3.64}$	$40.85_{\pm1.66}$	$59.96_{\pm0.91}$
	TransUNet [3]	All	✓		$96.32_{\pm0.19}$	$92.92_{\pm0.29}$	$85.04_{\pm0.15}$	$90.79_{\pm0.07}$	$94.08_{\pm0.07}$
		10	✓		$\mathbf{92.15}_{\pm0.40}$	$87.32_{\pm0.45}$	$66.51_{\pm2.28}$	$\mathbf{68.69}_{\pm0.58}$	$78.98_{\pm1.36}$
Automatic (adapted)	AutoSAM [17]	All	✓		$97.42_{\pm0.04}$	$93.59_{\pm0.03}$	$85.60_{\pm0.89}$	$89.57_{\pm0.54}$	$95.18_{\pm0.11}$
		10	✓		$90.64_{\pm1.84}$	$86.98_{\pm0.67}$	$67.09_{\pm4.82}$	$68.33_{\pm3.21}$	$\mathbf{84.17}_{\pm2.05}$
	PerSAM [22]	1	✓		$58.98_{\pm0.19}$	$36.13_{\pm0.00}$	$14.19_{\pm0.02}$	$27.64_{\pm9.48}$	$45.43_{\pm5.47}$
	Ours	All		✓	$92.88_{\pm1.27}$	$88.86_{\pm1.42}$	$79.82_{\pm0.74}$	$76.97_{\pm1.02}$	$86.91_{\pm2.08}$
		10		✓	$85.23_{\pm0.55}$	$\mathbf{88.38}_{\pm0.83}$	$\mathbf{73.56}_{\pm0.57}$	$58.96_{\pm2.28}$	$80.37_{\pm1.59}$

and AutoSAM [17] which uses a Harmonic Dense Net (41.6M parameters) to learn the prompt embedding. We train the UNet, TransUNet and AutoSAM on full segmentation masks with a standard Cross-entropy Dice loss. To improve the performance of the baseline models, longer training is used (200 epochs) with a larger batch size of 24 for TransUNet (following [3]). Similarly, the best results for PerSAM are obtained with the ViT-H backbone. Results for MedSAM are also included when prompted with the tightest bounding boxes (no noise).

3.3 Validation on Multiple Medical Segmentation Tasks

The Dice similarity score (DSC) is used as our evaluation criteria. We evaluate our approach on three datasets: CAMUS, an internal validation set of MedSAM, and HC18 and ACDC, two datasets never seen by MedSAM during training. This allows us to verify the ability of the prompt module to effectively learn which region to segment in both in-domain and out-of-domain data. Training is performed on both the entire training set and the difficult 10-shot regime.

Our results are given in Table 1 and Fig. 2. First, with only tight bounding box (BB) annotations, our approach trained on all samples is able to outperform a UNet trained on ground-truth segmentation masks for 2 different tasks (HC and LA). The most significant results are observed in the 10-shot setting. With only 1.3% (ACDC), 2% (HC18) and 2.9% (CAMUS) of the total training samples, our approach sees only a slight decrease in performance (except for RV segmentation) compared to the full-data setting. This contrasts with the considerable performance drop observed with UNet and TransUNet in multiple segmentation tasks, even when both methods are trained with ground-truth mask labels. Therefore, our prompt module-based approach is not only more computationally efficient to train than specialized models, but it also requires only weak labels and appears more robust in the few-shot setting. AutoSAM displays slightly better test dice scores than our approach, but AutoSAM requires full ground-truth masks and uses a much heavier model to learn the prompt, yielding a 3-fold increase in the training time. Additionally, PerSAM, which uses only one reference image, fails to generate convincing segmentations. Its poor performance on medical datasets may be due to the fact that PerSAM generates point prompts used by SAM, which are more likely to introduce ambiguity [12].

The benefits of our proposed methods are visually supported by Fig. 2. Our module trained on 10 training samples and tight bounding boxes yields segmentations much more convincing than those produced by a UNet trained on ground-truth masks. Given little training data, the UNet hallucinates large regions in the background. Our approach is also able to generate segmentation masks more faithful to the ground-truth than AutoSAM and PerSAM, two existing prompt-based adaptation methods for SAM.

4   Conclusion

This work proposes to automate a prompt-based universal model, such as Med-SAM, by generating task-specific prompt embeddings directly from the image embedding of the foundation model. Our add-on module that can be integrated directly into MedSAM removes its dependence on user inputs. More importantly, by applying tightness and size constraints, our module can be trained effectively with only bounding box annotations while keeping MedSAM frozen. Furthermore, our 10-shot experiments has demonstrated that the number of samples required to train the model could be considerably reduced without a substantial degradation of the model performance. By adding a lightweight prompt module that can be trained with only few weak labels, MedSAM can efficiently be automated for specific tasks with minimal annotation hurdles.

Acknowledgments. This work is supported by the Canada Research Chair on Shape Analysis in Medical Imaging, the Research Council of Canada (NSERC) and the Fonds de Recherche du Qubec – Nature et Technologies (FRQNT).

References

1. Bernard, O., Lalande, A., Zotti, C., Cervenansky, F., Yang, X., Heng, P.A., Cetin, I., Lekadir, K., Camara, O., Ballester, M.A.G., Sanroma, G., Napel, S., Petersen, S., Tziritas, G., Grinias, E., Khened, M., Kollerathu, V.A., Krishnamurthi, G., Rohé, M.M., Pennec, X., Sermesant, M., Isensee, F., Jäger, P., Maier-Hein, K.H., Full, P.M., Wolf, I., Engelhardt, S., Baumgartner, C.F., Koch, L.M., Wolterink, J.M., Iǎgum, I., Jang, Y., Hong, Y., Patravali, J., Jain, S., Humbert, O., Jodoin, P.M.: Deep Learning Techniques for Automatic MRI Cardiac Multi-Structures Segmentation and Diagnosis: Is the Problem Solved? IEEE Trans. Med. Imaging **37**(11), 2514–2525 (2018)
2. Brown, T., Mann, B., Ryder, N., Subbiah, M., Kaplan, J.D., Dhariwal, P., Neelakantan, A., Shyam, P., Sastry, G., Askell, A., Agarwal, S., Herbert-Voss, A., Krueger, G., Henighan, T., Child, R., Ramesh, A., Ziegler, D., Wu, J., Winter, C., Hesse, C., Chen, M., Sigler, E., Litwin, M., Gray, S., Chess, B., Clark, J., Berner, C., McCandlish, S., Radford, A., Sutskever, I., Amodei, D.: Language Models are Few-Shot Learners. In: Advances in Neural Information Processing Systems (NeurIPS). vol. 33, pp. 1877–1901 (2020)
3. Chen, J., Lu, Y., Yu, Q., Luo, X., Adeli, E., Wang, Y., Lu, L., Yuille, A.L., Zhou, Y.: TransUNet: Transformers Make Strong Encoders for Medical Image Segmentation (2021), http://arxiv.org/abs/2102.04306
4. Cheng, J., Ye, J., Deng, Z., Chen, J., Li, T., Wang, H., Su, Y., Huang, Z., Chen, J., Jiang, L., Sun, H., He, J., Zhang, S., Zhu, M., Qiao, Y.: SAM-Med2D (2023), http://arxiv.org/abs/2308.16184
5. Dosovitskiy, A., Beyer, L., Kolesnikov, A., Weissenborn, D., Zhai, X., Unterthiner, T., Dehghani, M., Minderer, M., Heigold, G., Gelly, S., Uszkoreit, J., Houlsby, N.: An Image is Worth 16x16 Words: Transformers for Image Recognition at Scale. In: International Conference on Learning Representations (ICLR) (2021)
6. Heuvel, T.L.A.v.d., Bruijn, D.d., Korte, C.L.d., Ginneken, B.v.: Automated measurement of fetal head circumference using 2D ultrasound images. Plos One **13**(8), e0200412 (2018)
7. Huang, Y., Yang, X., Liu, L., Zhou, H., Chang, A., Zhou, X., Chen, R., Yu, J., Chen, J., Chen, C., Liu, S., Chi, H., Hu, X., Yue, K., Li, L., Grau, V., Fan, D.P., Dong, F., Ni, D.: Segment anything model for medical images? Med. Image Anal. **92**, 103061 (2024)
8. Jia, M., Tang, L., Chen, B.C., Cardie, C., Belongie, S., Hariharan, B., Lim, S.N.: Visual Prompt Tuning. In: European Conference on Computer Vision (ECCV). vol. 13693, pp. 709–727 (2022)
9. Kervadec, H., Dolz, J., Wang, S., Granger, E., Ayed, I.B.: Bounding boxes for weakly supervised segmentation: Global constraints get close to full supervision. In: Conference on Medical Imaging with Deep Learning (MIDL). pp. 365–381 (2020)

10. Kirillov, A., Mintun, E., Ravi, N., Mao, H., Rolland, C., Gustafson, L., Xiao, T., Whitehead, S., Berg, A.C., Lo, W.Y., Dollár, P., Girshick, R.: Segment Anything. In: International Conference on Computer Vision (ICCV). pp. 3992–4003 (2023)
11. Leclerc, S., Smistad, E., Pedrosa, J., Østvik, A., Cervenansky, F., Espinosa, F., Espeland, T., Berg, E.A.R., Jodoin, P.M., Grenier, T., Lartizien, C., D'hooge, J., Lovstakken, L., Bernard, O.: Deep Learning for Segmentation Using an Open Large-Scale Dataset in 2D Echocardiography. IEEE Transactions on Medical Imaging **38**(9), 2198–2210 (2019)
12. Ma, J., He, Y., Li, F., Han, L., You, C., Wang, B.: Segment anything in medical images. Nat. Commun. **15**(1), 654 (2024)
13. Mazurowski, M.A., Dong, H., Gu, H., Yang, J., Konz, N., Zhang, Y.: Segment anything model for medical image analysis: An experimental study. Med. Image Anal. **89**, 102918 (2023)
14. Rajchl, M., Lee, M.C.H., Oktay, O., Kamnitsas, K., Passerat-Palmbach, J., Bai, W., Damodaram, M., Rutherford, M.A., Hajnal, J.V., Kainz, B., Rueckert, D.: DeepCut: Object Segmentation From Bounding Box Annotations Using Convolutional Neural Networks. IEEE Trans. Med. Imaging **36**(2), 674–683 (2017)
15. Ronneberger, O., Fischer, P., Brox, T.: U-Net: Convolutional Networks for Biomedical Image Segmentation. In: Medical Image Computing and Computer-Assisted Intervention (MICCAI). pp. 234–241 (2015)
16. Rother, C., Kolmogorov, V., Blake, A.: "GrabCut": interactive foreground extraction using iterated graph cuts. ACM Trans. Graph. **23**(3), 309–314 (2004)
17. Shaharabany, T.: AutoSAM: Adapting SAM to Medical Images by Overloading the Prompt Encoder. In: 34th British Machine Vision Conference (BMVC) (2023)
18. Wald, T., Roy, S., Koehler, G., Disch, N., Rokuss, M.R., Holzschuh, J., Zimmerer, D., Maier-Hein, K.: SAM.MD: Zero-shot medical image segmentation capabilities of the Segment Anything Model. In: Medical Imaging with Deep Learning (MIDL), short paper track (2023)
19. Wang, J., Xia, B.: Bounding Box Tightness Prior for Weakly Supervised Image Segmentation. In: Medical Image Computing and Computer Assisted Intervention (MICCAI). pp. 526–536 (2021)
20. Wu, Q., Zhang, Y., Elbatel, M.: Self-prompting Large Vision Models for Few-Shot Medical Image Segmentation. In: Domain Adaptation and Representation Transfer (MICCAI-DART) (2023)
21. Zhang, L., Deng, X., Lu, Y.: Segment Anything Model (SAM) for Medical Image Segmentation: A Preliminary Review. In: IEEE International Conference on Bioinformatics and Biomedicine (BIBM). pp. 4187–4194 (2023)
22. Zhang, R., Jiang, Z., Guo, Z., Yan, S., Pan, J., Ma, X., Dong, H., Gao, P., Li, H.: Personalize Segment Anything Model with One Shot. In: International Conference on Learning Representations (ICLR) (2024)

SAT-Morph: Unsupervised Deformable Medical Image Registration Using Vision Foundation Models with Anatomically Aware Text Prompt

Hao Xu[1], Tengfei Xue[1,2], Dongnan Liu[1], Fan Zhang[2,3], Carl-Fredrik Westin[2], Ron Kikinis[2], Lauren J. O'Donnell[2], and Weidong Cai[1(✉)]

[1] University of Sydney, Sydney, Australia
tom.cai@sydney.edu.au
[2] Harvard Medical School, Boston, USA
[3] University of Electronic Science and Technology of China, Chengdu, China

Abstract. Current unsupervised deformable medical image registration methods rely on image similarity measures. However, these methods are inherently limited by the difficulty of integrating important anatomy knowledge into registration. The development of vision foundation models (e.g., Segment Anything Model (SAM)) has attracted attention for their excellent image segmentation capabilities. Medical-based SAM aligns medical text knowledge with visual knowledge, enabling precise segmentation of organs. In this study, we propose a novel approach that leverages the vision foundation model to enhance medical image registration by integrating anatomical understanding of the vision foundation model into the medical image registration model. Specifically, we propose a novel unsupervised deformable medical image registration framework, called SAT-Morph, which includes Segment Anything with Text prompt (SAT) module and mask registration module. In the SAT module, the medical vision foundation model is utilized to segment anatomical regions within both moving and fixed images according to our designed text prompts. In the mask registration module, we take these segmentation results instead of traditionally used image pairs as the input of the registration model. Compared with utilizing image pairs as input, using segmentation mask pairs incorporates anatomical knowledge and improves the registration performance. Experiments demonstrate that SAT-Morph significantly outperforms existing state-of-the-art methods on both the Abdomen CT and ACDC cardiac MRI datasets. These results illustrate the effectiveness of integrating vision foundation models into medical image registration, showing the potential way for more accurate and anatomically-aware registration. Our code is available at https://github.com/HaoXu0507/SAT-Morph/.

Supplementary Information The online version contains supplementary material available at https://doi.org/10.1007/978-3-031-73471-7_8.

Z. Deng et al. (Eds.): MedAGI 2024, LNCS 15184, pp. 71–80, 2025.
https://doi.org/10.1007/978-3-031-73471-7_8

Keywords: Medical Image Registration · Vision Foundation Model · Text-prompted Segmentation · Multi-modal Learning

1   Introduction

Medical image registration refers to establishing the spatial correspondence between fixed images and moving images to maximize their similarity. Recently, many unsupervised deformable registration methods   [2–6,9,12,15,18,22,27] have emerged. TransMorph [5] combines the local capabilities of convolutional neural networks (CNN) [17] and the global capabilities of Transformer [23] to improve registration performance. TransMatch [6] further improves the effect by directly using the transformer's attention mechanism for image registration. DiffusionMorph [15] and its variant FSDiffReg [22] use the progressive denoising strategy of the diffusion model itself to perform progressive denoising and simultaneous registration CS-Reg [4] performs the cyclical self-training strategy to gradually refine pseudo labels.

However, these unsupervised deformable registration methods based on similarity measures can only equally weight the entire image but fail to allocate more weights to important anatomical regions. Therefore, these methods cannot integrate medical anatomy knowledge while performing registration. Currently, there is a great need for an unsupervised deformable image registration method to integrate the knowledge of medical anatomy.

The medical foundation model aligns medical text knowledge with medical image knowledge, showing promising results in various medical image tasks [7,10,11,14,19,25,26,28]. In particular, Segment Anything Model (SAM) [16] has recently attracted attention in the community because of its excellent image segmentation capabilities that only require simple prompts (box, point, or mask). Various variants of SAM [8,21,24,29] are constantly emerging to explore the boundaries of its capabilities, including medical image-based SAM [13]. Med-SAM [20] fine-tunes SAM and integrates medical knowledge in specific fields into the segmentation model, proving SAM's effectiveness in medical image registration. As the original SAM model is based on a 2D image architecture, it is not suitable for 3D medical image segmentation. SAM-MED3D [24] builds a medical image SAM model based on 3D images. Compared with the original SAM and SAM-MED2D [8], it achieves SOTA performance by using only 10 box prompts. However, the above variants of SAM based on medical images require the assistance of vision prompts (box, point, mask, etc.), which is still time-consuming and labor-intensive. Segment anything with text prompt (SAT) model aligns the textual knowledge and visual knowledge of the structure of medical images and achieves SOTA segmentation results by using only text prompts. Although medical-based SAM can deeply understand the various anatomical structures of medical images, there is no SAM-based method dedicated for medical image registration.

In this study, we propose a SAM-driven image registration framework called SAT-Morph, including a SAT module and a mask registration module. In the

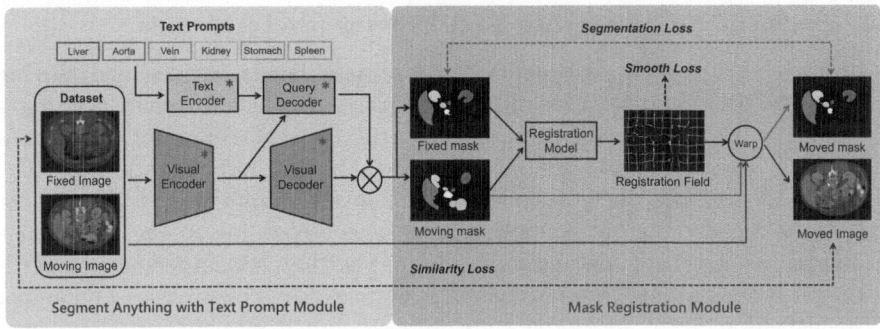

Fig. 1. The framework of SAT-Morph. Segment Anything with Text Prompt Module: Generating pseudo mask labels of image pairs according to our designed text prompts. Mask Registration Module: Taking pseudo mask labels as the input and output registration results. * denotes freezing model parameters.

SAT module, we use our designed text prompts to guide the powerful medical SAM model to segment paired registered images and generate pseudo mask labels of anatomical regions. In the mask registration module, instead of using moving and fixed image pairs as the input of the registration model, pseudo mask labels are utilized as the input to incorporate anatomical information and improve registration accuracy. To the best of our knowledge, we are the first to use pseudo mask labels as the input of registration model. We demonstrate the superiority of our proposed framework compared to the SOTA methods on the two datasets: ACDC Cardiac MRI and Abdomen CT datasets.

Our contributions are as follows. First, we propose a novel unsupervised deformable medical image registration method driven by the vision foundation model with our designed text prompts. Second, instead of using images as the input of the registration model, we utilize pseudo mask labels as the input, which integrates anatomical knowledge and improves registration performance. Third, our framework outperforms previous methods by a significant margin on ACDC cardiac MRI and Abdomen CT datasets. The framework demonstrates a potential way for more accurate and anatomically aware registration techniques.

2   Methodology

Our framework aims to obtain a spatial transform field U for register from M to F. In the SAT module, vision foundation model SAT generates pseudo masks by segmenting fixed and moving images into anatomical regions according to our designed text prompts. In the mask registration module, to reduce the training difficulty of the registration model and improve the registration performance, we utilize the pseudo mask labels of fixed and moving images as the input of the registration model.

2.1 Segment Anything with Text Prompt (SAT) Module

We utilize SAT module to generate pseudo mask labels of fixed and moving images. The SAT module is based on a pre-trained medical segmentation foundation model [29] with medical anatomical structure text as prompt, which consists of the visual encoder, visual decoder, text encoder, and query decoder. The text encoder and the query decoder take in text prompts and visual encoder and visual decoder take in medical image scans. According to the importance of organs, we generate text prompts for organs that are important for medical registration, and ignore unimportant organs. For example, for abdominal CT registration, liver and kidney are important, while colon and duodenum are unimportant. We fuse them to obtain segmentation masks according to our designed text prompts. Let a pair of fixed and moving images be $\{F, M\}$ and text prompts be $T = \{t1, ..., tn\}$:

$$F_{seg} = \theta_{SAT}(F, T), \tag{1}$$

where θ_{SAT} is the SAT segmentation model, and F_{seg} is the segmentation result of F. In the same way:

$$M_{seg} = \theta_{SAT}(M, T), \tag{2}$$

where M_{seg} is the segmentation result of M. Note that we design a set of text prompts for each dataset. Each set of text prompts can be used for all images of the entire dataset.

2.2 Mask Registration Module

We utilize the pair of $\{F_{seg}, M_{seg}\}$ as the pseudo labels of fixed and moving images. The registration model takes the pair of pseudo masks as input to compute the displacement field. Note that the field can be used as the spatial transform field not only from M_{seg} to F_{seg} but also from M to F. According to this property, we register the pseudo mask pair and the registration result can be directly applied to the image pair. Specifically, let the predicted displacement field be:

$$u = \theta_{Reg}(F_{seg}, M_{seg}), \tag{3}$$

where θ_{Reg} is the registration model. The spatial transform from the moving image to the moved image, and from the moving pseudo mask to the moved pseudo mask are as follows:

$$M = \phi(M, u) \tag{4}$$

and

$$M_{seg} = \phi(M_{seg}, u), \tag{5}$$

where M is the moved image, M_{seg} is the moved pseudo mask, and ϕ is the spatial transform function.

2.3 Loss Functions

Our loss function combines segmentation loss, similarity loss, and smooth loss.

Segmentation Loss. We calculate the segmentation loss between moved pseudo mask and fixed pseudo label to constrain the accuracy of the registration model. We take the combination of dice loss and focal loss as the segmentation loss, which is as follows:

$$L_{dice} = 1 - \frac{2\left|M'_{seg} \cap F_{seg}\right|}{\left|M'_{seg}\right| + \left|F_{seg}\right|}, \tag{6}$$

$$L_{focal} = \begin{cases} -F_{seg}(1 - M'_{seg})^\gamma \log M'_{seg}, & if \quad F_{seg} = 1, \\ (1 - F_{seg})(M'_{seg})^\gamma \log (1 - M'_{seg}), & otherwise. \end{cases} \tag{7}$$

$$L_{seg} = L_{dice} + \alpha L_{focal}, \tag{8}$$

where α is the hyper-parameter.

Similarity Loss. The mean squared error (MSE) loss between moved and fixed images is adopted as the similarity loss:

$$L_{sim} = \frac{1}{\Omega} \sum |M' - F|^2, \tag{9}$$

where Ω represents the image domain.

Smooth Loss. We utilize diffusion regularization on the spatial gradients of the deformable field as the smooth loss:

$$L_{smooth} = \sum_{p \in \Omega} || \nabla u(p)||^2, \tag{10}$$

where p denotes the voxel location. Finally, the overall loss function is as follows:

$$L = L_{seg} + \beta L_{sim} + \delta L_{smooth}, \tag{11}$$

where β and δ are the hyper-parameters.

3   Experiments and Results

3.1   Dataset and Text Prompts

Our methods are evaluated on two datasets of CT and MRI modalities: the Abdomen CT dataset and ACDC cardiac MRI dataset.

Abdomen CT Dataset. The Abdomen CT dataset contains 50 abdominal images. We randomly chose 40 images (780 pairs) for training and 10 images (45 pairs) for testing. The resolution is $192 \times 160 \times 256$ and each voxel size is $2 \times 2 \times 2$ mm. Our designed text prompts include spleen, right kidney, left kidney, gallbladder, esophagus, liver, stomach, aorta, inferior vena cava, portal vein and splenic vein, pancreas, right adrenal gland, and left adrenal gland.

ACDC Cardiac MRI Dataset. The ACDC cardiac MRI dataset contains 100 image pairs. We randomly chose 90 image pairs for training and 10 image pairs for testing, following the data split in previous works [15,22]. The resolution is $128 \times 128 \times 32$ and each voxel size is $1.5 \times 1.5 \times 3.15$ mm. Our designed text prompts include myocardium, left ventricle, and right ventricle.

3.2   Evaluation Metrics

We use dice score coefficient (DSC) and standard deviation of the Jacobian determinant (SDlogJ) as the evaluation metrics of the experiment, which two are widely used to evaluate on image registration [2,5,6,15,22]. A higher DSC indicates that the displacement field more accurately aligns the anatomy of relevant organs between the moving and fixed images. A lower SDlogJ indicates a smoother and more consistent displacement field between the moving and fixed images.

3.3   Implementation Details

We employ the trained SAT-nano model [29] as the segmentation model in our SAT module and freeze the model parameters throughout the training and inference stages. For the registration model, we choose TransMatch [6] as our basic registration model. Following the setup of TransMatch, our method is trained using Adam with a learning rate of 0.0004 and batch size 1 for 500 epochs. Regarding hyperparameters, α is 20, β is 1 , and δ is 0.04. The experiment is performed with Pytorch (v1.10) on an NVIDIA GeForce RTX 3090 GPU machine.

3.4   Comparison Experiments

Quantitative and Qualitative Comparisons on Abdomen CT Dataset. We compare our method with seven SOTA unsupervised deformable registration methods, including SyN [1], LDDMM [3], Deeds [12], VoxelMorph [2], Trans-Morph [5], TransMatch [6], and CS-Reg [4]. SyN, LDDMM, and Deeds are the traditional training-free registration methods and the others are deep learning based methods. As shown in Table 1, our method exceeds SOTA methods by a large margin. Specifically, our method achieves a margin of 14.73% and 11.77% higher DSC over Deeds and CS-Reg, respectively. To better demonstrate our results, we also show organ-specific results in the supplementary materials. It shows that our method achieves the best DSC in 12 out of 13 organs.

Figure 2 depicts qualitative comparisons of our approach against Deeds, CS-Reg, and SAT-Morph w/image. As seen, our approach achieves a more accurate registration result than competitors.

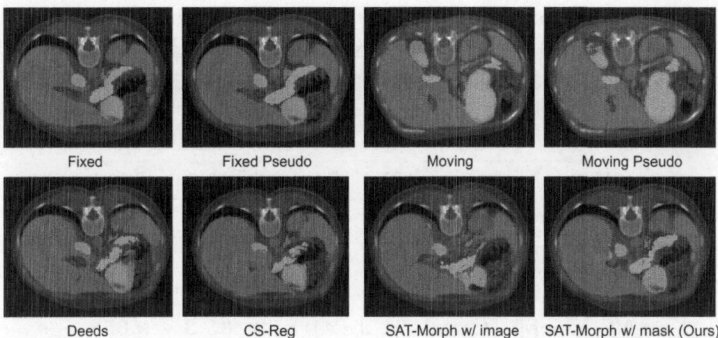

Fig. 2. Visualization of registration results for our proposed method and compared methods on Abdomen CT dataset.

Quantitative and Qualitative Comparisons on ACDC Cardiac MRI Dataset. We compare our method with four unsupervised deformable registration methods, including VoxelMorph, VoxelMorph-Diff [9], DiffuseMorph [15], and FSDiffReg [22]. As shown in Table 2, our method exceeds VoxelMorph, VoxelMorph-DIff, DiffuseMorph, FSDiffReg by 9.8%, 11.0%, 9.2%, and 6.6% DSC, respectively. In terms of specific anatomy regions, we lead the SOTA method by 11.0% and 9.4% DSC in the LV and Myo regions. Moreover, compared with other methods, our approach achieves the smallest SDlogJ. It shows that our method generates a smoother and more consistent displacement field.

Table 1. Quantitative comparisons on Abdomen CT dataset. ↑: higher is better, and ↓: lower is better.

	Method	DSC (%) ↑	SDlogJ ↓
Comparison	SyN [1]	23.25	N/A
	LDDMM [3]	25.51	N/A
	Deeds [12]	48.99	N/A
	VoxelMorph [2]	37.67	**0.143**
	TransMorph [5]	39.03	0.254
	TransMatch [6]	42.15	0.386
	CS-Reg [4]	51.95	0.149
Ablation Study	SAT-Morph w/image	59.39	0.974
	SAT-Morph w/mask (Ours)	**63.72**	0.910

The result of qualitative comparisons is shown in Fig. 3. As seen, our method has more complete and accurate registration results compared with other SOTA methods (e.g., FSDiffReg).

Table 2. Quantitative comparisons on ACDC Cardiac MRI dataset. ↑: higher is better, and ↓: lower is better. LV: left ventricle. Myo: myocardium. RV: right ventricle.

Method	DSC (%) ↑				SDlogJ ↓
	LV	Myo	RV	Overall	
VoxelMorph [2]	77.0	67.9	81.6	75.5	0.183
VoxelMorph-Diff [9]	75.5	65.9	81.5	74.3	0.182
DiffuseMorph [15]	78.3	67.8	82.1	76.1	0.178
FSDiffReg [22]	80.9	72.4	**82.7**	78.7	0.176
SAT-Morph (Ours)	**91.9**	**82.0**	81.9	**85.3**	**0.058**

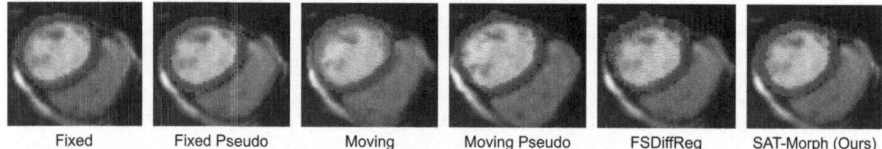

Fixed   Fixed Pseudo   Moving   Moving Pseudo   FSDiffReg   SAT-Morph (Ours)

Fig. 3. Visualization of registration results for our proposed method and compared methods on ACDC Cardiac MRI dataset.

Ablation Study. We compare the results of using the fixed and moving image pair or their pseudo masks as input of the registration model. As shown in Table 2, using pseudo masks as input exceeds using the image pair by 4.33% DSC. It demonstrates that using pseudo mask pairs can better train the model to achieve better registration accuracy than using image pairs as input of the registration model.

4   Conclusion

This work proposes SAT-Morph, a novel framework that leverages vision foundation model-driven approach into unsupervised deformable medical image registration. Our framework includes SAT module and mask segmentation module. The SAT module utilizes our uniquely designed text prompts to guide the vision foundation model in generating accurate pseudo mask labels. Then, these pseudo masks are taken as inputs by the registration model, replacing the traditional image pair inputs and thereby potentially pioneering a new direction in registration methodology. We demonstrate that using pseudo masks can achieve better registration accuracy than using image pairs as inputs to the registration model. We also show significant improvements in the Abdomen CT dataset and ACDC cardiac MRI dataset, highlighting its potential to set a new way for more accurate and anatomically-aware registration.

References

1. Avants, B.B., Epstein, C.L., Grossman, M., Gee, J.C.: Symmetric diffeomorphic image registration with cross-correlation: evaluating automated labeling of elderly and neurodegenerative brain. Med. Image Anal. **12**(1), 26–41 (2008)
2. Balakrishnan, G., Zhao, A., Sabuncu, M.R., Guttag, J., Dalca, A.V.: Voxelmorph: a learning framework for deformable medical image registration. IEEE Trans. Med. Imaging **38**(8), 1788–1800 (2019)
3. Beg, M.F., Miller, M.I., Trouvé, A., Younes, L.: Computing large deformation metric mappings via geodesic flows of diffeomorphisms. Int. J. Comput. Vision **61**, 139–157 (2005)
4. Bigalke, A., Hansen, L., Mok, T.C., Heinrich, M.P.: Unsupervised 3d registration through optimization-guided cyclical self-training. In: Greenspan, H., et al. (eds.) Medical Image Computing and Computer Assisted Intervention – MICCAI 2023. MICCAI 2023, LNCS, vol. 14229, pp. 677–687. Springer, Cham (2023). https://doi.org/10.1007/978-3-031-43999-5_64
5. Chen, J., Frey, E.C., He, Y., Segars, W.P., Li, Y., Du, Y.: Transmorph: transformer for unsupervised medical image registration. Med. Image Anal. **82**, 102615 (2022)
6. Chen, Z., Zheng, Y., Gee, J.C.: Transmatch: a transformer-based multilevel dual-stream feature matching network for unsupervised deformable image registration. IEEE Trans. Med. Imaging **43**(1), 15–27 (2024)
7. Chen, Z., et al.: Internvl: scaling up vision foundation models and aligning for generic visual-linguistic tasks. In: Proceedings of the IEEE/CVF Conference on Computer Vision and Pattern Recognition, pp. 24185–24198 (2024)
8. Cheng, J., et al.: Sam-med2d. arXiv preprint arXiv:2308.16184 (2023)
9. Dalca, A.V., Balakrishnan, G., Guttag, J., Sabuncu, M.R.: Unsupervised learning of probabilistic diffeomorphic registration for images and surfaces. Med. Image Anal. **57**, 226–236 (2019)
10. Gu, T., Liu, D., Li, Z., Cai, W.: Complex organ mask guided radiology report generation. In: Proceedings of the IEEE/CVF Winter Conference on Applications of Computer Vision, pp. 7995–8004 (2024)
11. Gu, T., Yang, K., Liu, D., Cai, W.: Lapa: latent prompt assist model for medical visual question answering. In: Proceedings of the IEEE/CVF Conference on Computer Vision and Pattern Recognition (CVPR) Workshops, pp. 4971–4980 (June 2024)
12. Heinrich, M.P., Maier, O., Handels, H.: Multi-modal multi-atlas segmentation using discrete optimisation and self-similarities. VISCERAL Challenge@ ISBI **1390**, 27 (2015)
13. Huang, Y., et al.: Segment anything model for medical images? Med. Image Anal. **92**, 103061 (2024)
14. Jin, H., Che, H., Lin, Y., Chen, H.: Promptmrg: diagnosis-driven prompts for medical report generation. In: Proceedings of the AAAI Conference on Artificial Intelligence, vol. 38, pp. 2607–2615 (2024)
15. Kim, B., Han, I., Ye, J.C.: Diffusemorph: unsupervised deformable image registration using diffusion model. In: Avidan, S., Brostow, G., Cissé, M., Farinella, G.M., Hassner, T. (eds.) Computer Vision – ECCV 2022. ECCV 2022. LNCS, vol. 13691. Springer, Cham (2022). https://doi.org/10.1007/978-3-031-19821-2_20
16. Kirillov, A., et al.: Segment anything. arXiv preprint arXiv:2304.02643 (2023)
17. LeCun, Y., Bengio, Y., et al.: Convolutional networks for images, speech, and time series. The Handbook of Brain Theory and Neural Networks, vol. 3361, issue 10, p. 1995 (1995)

18. Li, Z., et al.: Samconvex: fast discrete optimization for CT registration using self-supervised anatomical embedding and correlation pyramid. In: Greenspan, H., et al. (eds.) Medical Image Computing and Computer Assisted Intervention – MIC-CAI 2023. MICCAI 2023. LNCS, vol. 14229. Springer, Cham (2023). https://doi.org/10.1007/978-3-031-43999-5_53

19. Liao, W., et al.: Differentiating chatgpt-generated and human-written medical texts: quantitative study. JMIR Med. Educ. 9(1), e48904 (2023)

20. Ma, J., He, Y., Li, F., Han, L., You, C., Wang, B.: Segment anything in medical images. Nat. Commun. 15(1), 654 (2024)

21. Mazurowski, M.A., Dong, H., Gu, H., Yang, J., Konz, N., Zhang, Y.: Segment anything model for medical image analysis: an experimental study. Med. Image Anal. 89, 102918 (2023)

22. Qin, Y., Li, X.: Fsdiffreg: feature-wise and score-wise diffusion-guided unsupervised deformable image registration for cardiac images. In: Greenspan, H., et al. (eds.) Medical Image Computing and Computer Assisted Intervention – MICCAI 2023. MICCAI 2023. LNCS, vol. 14229. Springer, Cham. (2023). https://doi.org/10.1007/978-3-031-43999-5_62

23. Vaswani, A., et al.: Attention is all you need. Adv. Neural Inform. Process. Syst. 30 (2017)

24. Wang, H., et al.: Sam-med3d. arXiv preprint arXiv:2310.15161 (2023)

25. Wang, W., et al.: Visionllm: large language model is also an open-ended decoder for vision-centric tasks. Adv. Neural Inform. Process. Syst. 36 (2024)

26. Xu, J., et al.: Data set and benchmark (medgpteval) to evaluate responses from large language models in medicine: evaluation development and validation. JMIR Med. Inform. 12(1), e57674 (2024)

27. Zhang, F., Wells, W.M., O'Donnell, L.J.: Deep diffusion Mri registration (ddmreg): a deep learning method for diffusion MRI registration. IEEE Trans. Med. Imaging 41(6), 1454–1467 (2021)

28. Zhang, S., Metaxas, D.: On the challenges and perspectives of foundation models for medical image analysis. Med. Image Anal. 102996 (2023)

29. Zhao, Z., et al.: One model to rule them all: towards universal segmentation for medical images with text prompts. arXiv preprint arXiv:2312.17183 (2023)

Promptable Counterfactual Diffusion Model for Unified Brain Tumor Segmentation and Generation with MRIs

Yiqing Shen, Guannan He, and Mathias Unberath[✉]

Johns Hopkins University, Baltimore, MD 21218, USA
{yshen92,unberath}@jhu.edu

Abstract. Brain tumor analysis in Magnetic Resonance Imaging (MRI) is crucial for accurate diagnosis and treatment planning. However, the task remains challenging due to the complexity and variability of tumor appearances, as well as the scarcity of labeled data. Traditional approaches often address tumor segmentation and image generation separately, limiting their effectiveness in capturing the intricate relationships between healthy and pathological tissue structures. We introduce a novel promptable counterfactual diffusion model as a unified solution for brain tumor segmentation and generation in MRI. The key innovation lies in our mask-level prompting mechanism at the sampling stage, which enables guided generation and manipulation of specific healthy or unhealthy regions in MRI images. Specifically, the model's architecture allows for bidirectional inference, which can segment tumors in existing images and generate realistic tumor structures in healthy brain scans. Furthermore, we present a two-step approach for tumor generation and position transfer, showcasing the model's versatility in synthesizing realistic tumor structures. Experiments on the BRATS2021 dataset demonstrate that our method outperforms traditional counterfactual diffusion approaches [17], achieving a mean IoU of 0.653 and mean Dice score of 0.785 for tumor segmentation, outperforming the 0.344 and 0.475 of conventional counterfactual diffusion model. Our work contributes to improving brain tumor detection and segmentation accuracy, with potential implications for data augmentation and clinical decision support in neuro-oncology. The code is available at https://github.com/arcadelab/counterfactual_diffusion.

Keywords: Counterfactual Diffusion Model · Deep Learning · MRI · Tumor Segmentation

1 Introduction

Brain tumor analysis in Magnetic Resonance Imaging (MRI) is an important task in medical diagnosis and treatment planning [4]. The accurate segmentation

---

Yiqing Shen, Guannan He: Equal contributions

© The Author(s), under exclusive license to Springer Nature Switzerland AG 2025
Z. Deng et al. (Eds.): MedAGI 2024, LNCS 15184, pp. 81–90, 2025.
https://doi.org/10.1007/978-3-031-73471-7_9

and characterization of tumors from MRI scans are essential for effective patient care in neuro-oncology [8]. However, this task remains challenging due to the complexity and variability of tumor appearances across different MRI modalities and the need for precise delineation of tumor boundaries. Recent advancements in deep learning, particularly in the fields of computer vision and medical image analysis, have shown promising results in addressing these challenges by training an end-to-end model like UNet [5]. Diffusion models, which have demonstrated remarkable capabilities in image generation tasks, offer a new perspective on medical image analysis by unifying discriminative and generative tasks [7].

Despite these advancements, several key challenges persist in brain tumor analysis. High-quality, annotated MRI datasets for brain tumors are scarce, hindering the development of robust segmentation models. The diverse appearance of brain tumors across patients and MRI modalities makes it difficult for traditional segmentation models to generalize effectively. Many existing deep learning models operate as "black boxes", making it challenging for clinicians to understand and trust their decisions. Furthermore, current approaches often struggle to generate realistic tumor images or manipulate tumor characteristics in a controlled manner. To address these challenges, we innovatively propose a unified approach that combines the strengths of diffusion models, counterfactual reasoning, and prompt-guided generation. This combination offers several advantages. Counterfactual reasoning allows the model to learn the relationship between healthy and pathological tissue, potentially improving segmentation accuracy and generalization. Prompt-guided generation enables more controlled and interpretable tumor analysis, allowing clinicians to interact with the model and explore various scenarios. The unified framework for both segmentation and generation can leverage the synergies between these tasks, potentially improving performance on both fronts. For segmentation, our approach aims to accurately identify and delineate tumor regions across multiple MRI modalities (T1, T1-ce, T2, and FLAIR). In terms of tumor generation, we focus on creating synthetic tumor images and achieving tumor position transfer, which has implications for data augmentation and hypothetical disease progression modeling.

The key contributions of our work are three-fold. Firstly, we present a promptable counterfactual diffusion sampling method that outperforms traditional approaches in both tumor segmentation and generation tasks by enabling manual intervention in the diffusion denoising process. Secondly, we provide a comprehensive comparison of Transformer-based and UNet-based denoising architectures in the context of the counterfactual diffusion model. We show that Transformer can capture global context more effectively than traditional convolutional approaches, leading to improved performance in handling complex tumor morphologies. We also introduce a two-step tumor generation approach, demonstrating the versatility of our method in manipulating and synthesizing realistic tumor structures.

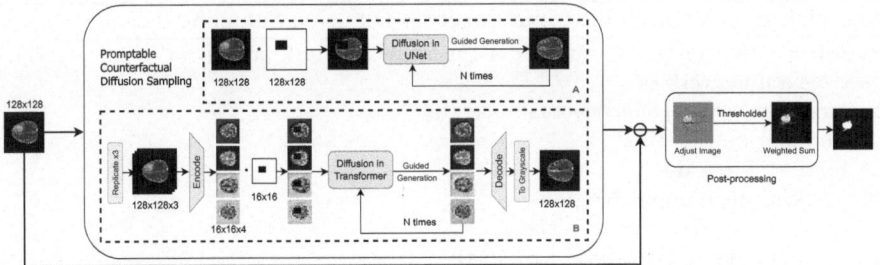

Fig. 1. The overall pipeline of the promptable counterfactual diffusion sampling scheme. (A) UNet-based approach: A 128×128 input image is combined with a binary mask prompt for guided generation through a UNet-based denoising network, iterated N times. (B) Transformer-based approach: The input image is replicated, encoded, and processed at multiple resolutions ($128 \times 128 \times 3$, $16 \times 16 \times 4$) before undergoing guided generation through a Transformer-based denoising network, also iterated N times. Both approaches leverage mask-guided diffusion for region-specific generation. The final output undergoes post-processing, including thresholding and weighted sum computation, to produce the segmented tumor region.

2   Methods

In this section, we detail the architecture of our promptable counterfactual diffusion model. We further enhance the diffusion process for the counterfactual diffusion model [17] by replacing the conventional U-Net denoising network [6] backbone with a Transformer-based denoising network [13] to show its improvement is also applicable beyond DDPM. The overall framework is summarized in Fig. 1.

Promptable Counterfactual Diffusion Sampling. Our proposed promptable counterfactual diffusion sampling scheme builds upon the DDPM [6]. It enables both prompt-guided tumor segmentation [1] and generation within a unified diffusion model by formulating them uniformly as a counterfactual diffusion process. Inspired by previous work [10], our promptable counterfactual diffusion sampling scheme extends this approach to allow for the generation of specific areas in MRI images as either healthy or unhealthy, based on a given mask prompt. During the reverse diffusion process, we incorporate mask prompts to guide the generation of desired regions, ensuring semantic consistency with the rest of the MR image. By integrating this prompting mechanism with DDPM's iterative denoising process, we can produce high-quality, diverse counterfactual outputs for both healthy and tumor regions. This flexibility enables the use of a single model for both segmentation and augmentation in MRI analysis, streamlining the workflow and potentially improving overall performance. The complete process of our promptable counterfactual diffusion sampling scheme is detailed in Algorithm 1 and illustrated.

Algorithm 1. Promptable Counterfactual Diffusion Sampling Scheme

---

Require: Clean image x_0, number of timesteps T, mask prompt m, guidance signal c, neural network ϵ_θ

Ensure: Generated counterfactual image x_0^{gen}

1: Initialize $x_T \sim \mathcal{N}(0, \mathbf{I})$
2: **for** $t = T$ to 1 **do**
3:     **Sample Known Regions:**
4:     $x_{t-1}^{\text{known}} \sim q(x_{t-1}|x_t, x_0)$ where
5:     $q(x_{t-1}|x_t, x_0) = \mathcal{N}(\tilde{\mu}_t(x_t, x_0), \tilde{\beta}_t \mathbf{I})$
6:     $\tilde{\mu}_t(x_t, x_0) = \frac{\sqrt{\bar{\alpha}_{t-1}}\beta_t}{1-\bar{\alpha}_t} x_0 + \frac{\sqrt{\alpha_t}(1-\bar{\alpha}_{t-1})}{1-\bar{\alpha}_t} x_t$
7:     **Guided Sampling for Unknown Regions:**
8:     Predict noise: $\hat{\epsilon} = \epsilon_\theta(x_t, t)$
9:     Compute denoised estimate: $x_0^t = \frac{x_t - \sqrt{1-\bar{\alpha}_t}\hat{\epsilon}}{\sqrt{\bar{\alpha}_t}}$
10:    Apply guidance: $x_0^t \leftarrow x_0^t + c \cdot \nabla_{x_0^t} \log p(x_0^t)$
11:    $x_{t-1}^{\text{guided}} \sim \mathcal{N}(\mu_\theta(x_t, t, x_0^t), \Sigma_\theta(x_t, t))$
12:    **Combine Samples:**
13:    $x_{t-1} = m \odot x_{t-1}^{\text{known}} + (1-m) \odot x_{t-1}^{\text{guided}}$
14: **end for**
15: **Output:** $x_0^{\text{gen}} \leftarrow x_0$

Transformer-Based Denoising Network. We further adopt the Diffusion Transformer (DiT) [13] for the promptable counterfactual diffusion sampling scheme in brain tumor segmentation and generation from MRI. Specifically, our approach replaces the traditional U-Net denoising network [16] with a Transformer-based backbone, leveraging the superior global context capture capabilities of Transformers [18] to enhance the diffusion model's performance in generating high-quality MRI images. Building on the foundational concepts of DiT, we introduce several key enhancements tailored to MRI. Firstly, we integrate a prompting mechanism for guided tumor segmentation and generation, enabling the creation of specific healthy or unhealthy regions in MRI images based on mask inputs and user-defined prompts. This exploits the Transformers' global context awareness to achieve more precise and contextually relevant image generation. Second, the preprocessing of input MRI data involves utilizing a pre-trained variational autoencoder (VAE) [9] from Stable Diffusion [15] to encode input images into a latent space. These latent representations are then transformed into token sequences through patchification. Our processing pipeline processes these token sequences through multiple DiT blocks. Each DiT block incorporates in-context conditioning, cross-attention, and adaptive layer normalization (adaLN), similar to adaptive normalization layers in GANs [14]. These components collectively improve the denoising network's ability to manage conditioning information, enhancing both performance and scalability. Finally, in this work, we implement the DiT B-2 architecture size, as a balance between computational efficiency and model performance. Finally, to address the RGB input requirement of the pre-trained VAE, we process each MRI modality separately by replicating it across three channels, creating an RGB-like input.

Initial Segmentation Derivation. The initial segmentation is obtained by comparing the original MRI image (factual) with the generated healthy version (counterfactual) produced by our model. Specifically, we compute the difference image between these two:

$$D(x, y) = |I_f(x, y) - I_c(x, y)|,\tag{1}$$

where $I_f(x, y)$ is the intensity of the factual image at location (x, y), and $I_c(x, y)$ is the intensity of the counterfactual image at the same location. This difference image $D(x, y)$ highlights areas where the model has made changes, potentially indicating tumor regions. The resulting difference image serves as our preliminary segmentation. However, this raw output may contain noise or ambiguities, necessitating further refinement through post-processing.

Post-processing Refinement for Segmentation. After obtaining preliminary tumor segmentation results using the proposed promptable counterfactual diffusion sampling scheme, we apply a series of post-processing techniques to refine these results, ensuring higher accuracy and robustness. Our post-processing pipeline consists of three main steps: contrast and brightness adjustment, thresholding, and channel-weighted sum computation.

First, we enhance the initial segmentation output, which represents the difference image between the ground truth and the generated healthy image, by adjusting its contrast and brightness. This step amplifies the contrast and highlights tumor regions more clearly, facilitating easier distinction. Next, we convert the contrast-adjusted image into a binary image using Otsu's thresholding method [12] from the `skimage.filters` module. It determines the optimal threshold value to effectively separate tumor regions from the background. Finally, we process each of the four channels corresponding to the four MRI modalities separately. We assign an equal weight of 0.25 to each modality's result and integrate them by summing the weighted outputs. A voxel is classified as having a tumor if the cumulative weight is 0.5 or more, ensuring that tumor presence is determined based on majority overlap across channels. This process is formalized as

$$S(x, y) = \sum_{i=1}^{4} w_i \cdot M_i(x, y)\tag{2}$$

where $w_i = 0.25$ for each modality i, and $M_i(x, y)$ represents the binary mask for modality i at location (x, y). A tumor is identified at location (x, y) if $S(x, y) \geq 0.5$. It ensures that a location is classified as a tumor if half or more of the modalities indicate the presence of a tumor.

3   Results

Datasets. Our model training and experiments utilized the BRATS2021 (Brain Tumor Segmentation Challenge 2021) dataset [2,3,11], a well-established MRI benchmark for brain tumor segmentation tasks. This dataset comprises multi-modal MRI scans from patients, featuring four MRI sequences per patient: T1-weighted (T1), post-contrast T1-weighted (T1ce), T2-weighted (T2), and Fluid Attenuated Inversion Recovery (FLAIR).

Implementation Details. We employed data from 1,000 patients for training, with each patient contributing 155 2D slices per modality, where we randomly split 104 slices as the test set. All implementations are made on the 2D slice level. For each slice, we concatenated the four modalities along the channel dimension into a single file. These images were then resized to 128×128 pixels and normalized. In our training process, we classified axial slices containing at least one tumor pixel as "unhealthy," while those without any tumor pixels were considered "healthy". These two categories of data were used to train the conditional diffusion model. We conducted the training process on a single NVIDIA RTX A6000 48 GB GPU, optimizing our model with a learning rate of 1e-4, the AdamW optimizer, a batch size of 128, and training for 15,000 steps. We assessed the performance of tumor segmentation using two primary metrics: (i) the Dice coefficient and (ii) Intersection over Union (IoU). To evaluate our promptable counterfactual diffusion sampling method, we compared it with the traditional counterfactual sampling method [17]. We assessed four methods in total: (i) Promptable counterfactual diffusion sampling with a Transformer denoising network (our proposal), (ii) Promptable counterfactual diffusion sampling with a UNet denoising network (our proposed method, but as ablation), (iii) Counterfactual sampling with a Transformer denoising network (*i.e.*, without mask prompt, baseline), and (iv) Counterfactual sampling with a UNet denoising network (*i.e.*, without mask prompt, baseline).

Table 1. Comparison of brain tumor segmentation performance on the BRATS2021 dataset. We compare against promptable vs. traditional counterfactual diffusion model [17] with Transformer and UNet as denoising networks.

Method	Mean IoU	Mean Dice
Promptable Counterfactual w/Transformer	**0.653**	**0.785**
Promptable Counterfactual w/UNet	0.647	0.772
Counterfactual Diffusion w/Transformer	0.366	0.479
Counterfactual Diffusion w/UNet	0.344	0.475

Evaluation of Tumor Segmentation. Our assessment encompassed four distinct methods: Promptable Counterfactual with Transformer (PCF+ Transformer), Promptable Counterfactual with UNet (PCF+UNet), Counterfactual Diffusion Sampling with Transformer (CF+Transformer), and Counterfactual Diffusion Sampling with UNet (CF+UNet) [17]. The quantitative

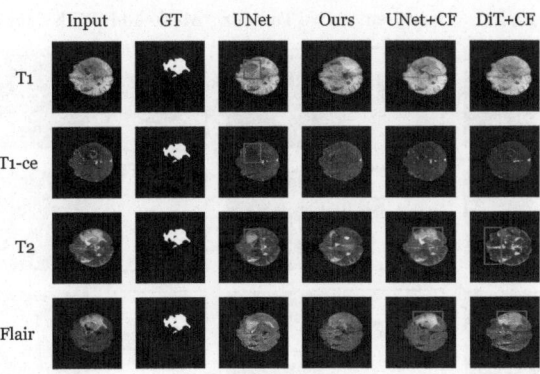

Fig. 2. Comparative visualization of brain tumor segmentation results across four MRI modalities (T1, T1-ce, T2, and Flair). The columns show: Input (original MRI), GT (ground truth), UNet (our PCF+UNet), Ours (our PCF+Transformer), UNet+CF (Counterfactual Diffusion with UNet), and DiT+CF (Counterfactual Diffusion with Transformer). Red boxes highlight areas where methods struggled to accurately segment or remove tumors. Our PCF+Transformer method (Ours) demonstrates superior tumor removal across all modalities.

results of this evaluation are presented in Table 1. Our proposed method (*i.e.*, PCF+Transformer) demonstrated superior performance, achieving the highest Mean IoU of 0.653 and Mean Dice score of 0.785. This outstanding performance underscores the effectiveness of our approach in accurate tumor segmentation. Notably, both promptable counterfactual methods (PCF+Transformer and PCF+UNet) consistently outperformed their traditional counterfactual diffusion [17] counterparts, highlighting the significant advantages of guided generation in medical imaging tasks. The visual examples of our segmentation results, as illustrated in Fig. 2, provide further insight into the performance of each method. PCF+Transformer exhibited the most impressive results, effectively identifying and delineating tumor regions across all MRI modalities. While PCF+UNet also showed strong performance, it occasionally fell short of completely delineating tumor boundaries. In contrast, the CF methods demonstrated inconsistent performance across different MRI modalities, with some instances of incomplete tumor identification or false positives. These results collectively emphasize the superiority of our promptable counterfactual approach, particularly when coupled with a Transformer denoising network.

Evaluation of Tumor Generation. We developed a two-step approach to generate tumor images and achieve tumor position transfer, leveraging the capabilities of our promptable counterfactual model. The first step is tumor removal, where we utilize our promptable counterfactual model to transform unhealthy slices containing tumors into healthy slices. This process involved the precise erasure of tumor regions while preserving the integrity of surrounding brain tissues. It demonstrates the model's ability to comprehend and manipulate complex

Input   Mask-Remove No Tumour   Mask-Add With Tumour

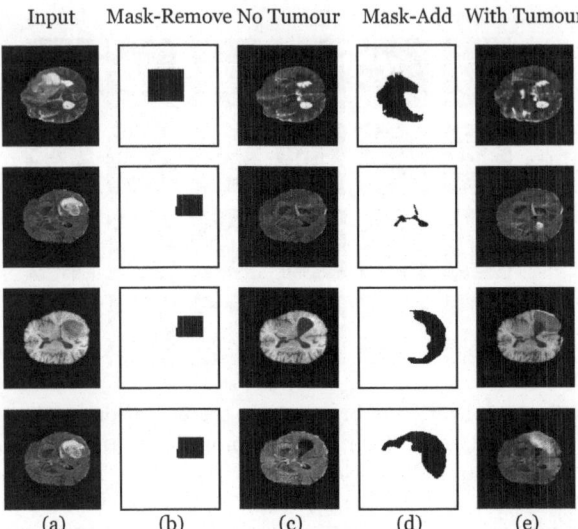

(a)   (b)   (c)   (d)   (e)

Fig. 3. Visualization of the tumor regeneration process: (a) Original unhealthy slice, (b) Mask for removing the tumor, (c) Healthy slice after tumor removal, (d) Random mask for tumor generation (e) Regenerated tumor in a new location.

anatomical structures within MRI images. The second step is tumor regeneration, which employs randomly generated masks to guide the regeneration of tumors within the previously "healed" slices. By placing these masks, we were able to generate new tumor regions in different locations, effectively achieving tumor position transfer. This capability showcases the model's flexibility in synthesizing realistic tumor structures while maintaining anatomical plausibility. The results of this two-step process are visually demonstrated in Fig. 3. This figure illustrates the progression from original tumor-containing images, through the tumor removal phase, to the final images with regenerated tumors in new positions. This approach not only demonstrates the versatility of our promptable counterfactual model but also opens up new possibilities for data augmentation. By generating diverse tumor presentations, we can potentially provide valuable training data for medical professionals.

4   Conclusion

In this paper, we introduced a promptable counterfactual diffusion sampling method, which demonstrates the unification of tumor segmentation and generation tasks. Our experiments revealed important distinctions between the UNet and DiT denoising network backbones. The UNet-based model, while effective in many scenarios, occasionally generated black regions when presented with large masks or those extending beyond the brain shape in the MRI. In contrast, the DiT backbone, operating in the latent space, demonstrated superior robustness and stability in handling these complex cases. Despite the lower resolution resulting from latent space processing, the DiT backbone excelled in managing intricate masks more effectively than its UNet counterpart. Our approach can remove the tumor while maintaining the integrity of the original image structure, representing a significant improvement in the field. Future work could explore its application to other medical imaging domains and investigate its potential in clinical decision support systems.

References

1. Amit, T., Shaharbany, T., Nachmani, E., Wolf, L.: Segdiff: image segmentation with diffusion probabilistic models. arXiv preprint arXiv:2112.00390 (2021)
2. Baid, U., et al.: The rsna-asnr-miccai brats 2021 benchmark on brain tumor segmentation and radiogenomic classification. arXiv preprint arXiv:2107.02314 (2021)
3. Bakas, S., et al.: Advancing the cancer genome atlas glioma MRI collections with expert segmentation labels and radiomic features. Sci. Data 4(1), 1–13 (2017)
4. Cè, M., et al..: Artificial intelligence in brain tumor imaging: a step toward personalized medicine. Curr. Oncol. 30(3), 2673–2701 (2023)
5. Conze, P.H., Andrade-Miranda, G., Singh, V.K., Jaouen, V., Visvikis, D.: Current and emerging trends in medical image segmentation with deep learning. IEEE Trans. Radiat. Plasma Med. Sci. 7(6), 545–569 (2023)
6. Ho, J., Jain, A., Abbeel, P.: Denoising diffusion probabilistic models. Adv. Neural. Inf. Process. Syst. 33, 6840–6851 (2020)
7. Kazerouni, A., et al.: Diffusion models in medical imaging: a comprehensive survey. Med. Image Anal. 88, 102846 (2023)
8. Khalighi, S., Reddy, K., Midya, A., Pandav, K.B., Madabhushi, A., Abedaltha-gafi, M.: Artificial intelligence in neuro-oncology: advances and challenges in brain tumor diagnosis, prognosis, and precision treatment. NPJ Precision Oncol. 8(1), 80 (2024)
9. Kingma, D.P., Welling, M.: Auto-encoding variational bayes. arXiv preprint arXiv:1312.6114 (2013)
10. Lugmayr, A., Danelljan, M., Romero, A., Yu, F., Timofte, R., Van Gool, L.: Repaint: Inpainting using denoising diffusion probabilistic models. In: Proceedings of the IEEE/CVF Conference on Computer Vision and Pattern Recognition, pp. 11461–11471 (2022)
11. Menze, B.H., Jakab, A., Bauer, S., Kalpathy-Cramer, J., et al.: The multimodal brain tumor image segmentation benchmark (brats). IEEE Trans. Med. Imaging 34(10), 1993–2024 (2015). https://doi.org/10.1109/TMI.2014.2377694
12. Otsu, N.: A threshold selection method from gray-level histograms. IEEE Trans. Syst. Man Cybern. 9(1), 62–66 (1979). https://doi.org/10.1109/TSMC.1979.4310076
13. Peebles, W., Xie, S.: Scalable diffusion models with transformers. In: Proceedings of the IEEE/CVF International Conference on Computer Vision, pp. 4195–4205 (2023)
14. Perez, E., Strub, F., De Vries, H., Dumoulin, V., Courville, A.: Film: Visual reasoning with a general conditioning layer. In: Proceedings of the AAAI conference on Artificial Intelligence, vol. 32 (2018)

15. Rombach, R., Blattmann, A., Lorenz, D., Esser, P., Ommer, B.: High-resolution image synthesis with latent diffusion models. In: Proceedings of the IEEE/CVF Conference on Computer Vision and Pattern Recognition, pp. 10684–10695 (2022)
16. Ronneberger, O., Fischer, P., Brox, T.: U-net: Convolutional networks for biomedical image segmentation. In: Navab, N., Hornegger, J., Wells, W., Frangi, A. (eds.) Medical Image Computing and Computer-Assisted Intervention – MICCAI 2015. MICCAI 2015. LNCS, vol. 9351. Springer, Cham (2015). https://doi.org/10.1007/978-3-319-24574-4_28
17. Sanchez, P., Kascenas, A., Liu, X., O'Neil, A.Q., Tsaftaris, S.A.: What is healthy? generative counterfactual diffusion for lesion localization. In: Mukhopadhyay, A., Oksuz, I., Engelhardt, S., Zhu, D., Yuan, Y. (eds.) Deep Generative Models. DGM4MICCAI 2022. LNCS, vol. 13609. Springer, Cham (2022). https://doi.org/10.1007/978-3-031-18576-2_4
18. Vaswani, A., et al.: Attention is all you need. Adv. Neural Inform. Process. Syst. **30** (2017)

D-Rax: Domain-Specific Radiologic Assistant Leveraging Multi-modal Data and eXpert Model Predictions

Hareem Nisar[1], Syed Muhammad Anwar[1,2(✉)], Zhifan Jiang[1],
Abhijeet Parida[1,3], Ramon Sanchez-Jacob[1,2], Vishwesh Nath[4],
Holger R. Roth[4], and Marius George Linguraru[1,2]

[1] Children's National Hospital, Washington, DC, USA
sanwar@childrensnational.org
[2] George Washington University, Washington, DC, USA
[3] Universidad Politécnica de Madrid, Madrid, Spain
[4] Nvidia Corporation, Santa Clara, CA, USA

Abstract. Large vision language models (VLMs) have progressed incredibly from research to applicability for general-purpose use cases. LLaVA-Med, a pioneering large language and vision assistant for biomedicine, can perform multi-modal biomedical image and data analysis to provide a natural language interface for radiologists. While it is highly generalizable and works with multi-modal data, it is currently limited by well-known challenges in the large language model space. Hallucinations and imprecision in responses can lead to misdiagnosis, which currently hinders VLMs' clinical adaptability. To create precise, user-friendly models in healthcare, we propose D-Rax- a domain-specific, conversational, radiologic assistance tool that can be used to gain insights about a particular radiologic image. In this study, we enhance the conversational analysis of chest X-ray (CXR) images to support radiological reporting, offering comprehensive insights from medical imaging and aiding in the formulation of accurate diagnosis. D-Rax is achieved by fine-tuning the LLaVA-Med architecture on our curated enhanced instruction-following data, comprising of images, instructions, as well as disease diagnosis and demographic predictions derived from MIMIC-CXR imaging data, CXR-related visual question answer (VQA) pairs, and predictive outcomes from multiple expert AI models. We observe statistically significant improvement in responses when evaluated for both open and close-ended conversations. Leveraging the power of state-of-the-art diagnostic models combined with VLMs, D-Rax empowers clinicians to interact with medical images using natural language, which could potentially streamline their decision-making process, enhance diagnostic accuracy, and conserve their time.

Keywords: Large vision language models · Radiologic assistant · Chest X-ray · Expert models

H. Nisar, S. M. Anwar and Z. Jiang—These authors contributed equally.

1   Introduction

Burnout in radiology is on the rise globally leading to chronic job dissatisfaction and critical under-staffing [4]. Radiologists routinely spend extensive time meticulously analyzing medical images to identify pathologies and diagnose diseases, which is vital in guiding treatment decisions and ensuring appropriate patient care. The retrospective error rate among radiologic exams has been reported to be around 30% [17]. Cindy et al. [17] assess these errors to be either cognitive, like false initial assessment, framing bias (i.e., misinterpretation caused by choice of words), and premature closure of a case, or system-related errors, such as long working shifts, repetitive tasks, and lighting conditions. Many of these factors contribute to visual and mental fatigue for radiologists, further contributing to misdiagnosis and poor patient outcomes. Another challenge is miscommunication between radiologists, clinicians, and patients, often caused by inefficient reporting.

With the constant increase in workload in radiology departments [2], generative artificial intelligence (AI) can play a crucial role in reducing the burden and improving healthcare [21]. Recent large vision language models (VLMs) such as LLaVA-Med [18] have been created to assist clinicians in interpreting complex medical imaging and provide visual question answering (VQA) in natural language settings. Despite its enhanced capabilities for medical image analysis and interpretation, LLaVA-Med is highly generalized and cannot precisely answer specific questions [25], as well as suffers from hallucinations that can result in misdiagnosis. Another challenge in the integration and adoption of AI-driven technologies among healthcare professionals is user-friendliness of the tool [8]. These clinical and technological challenges necessitate a "Radiology Assistant" tool that can facilitate report writing and provide a natural-language interface to discuss imaging features, pathological findings, and disease diagnosis with the radiologist.

To address these challenges, we propose a novel, domain-specific VLM, called D-Rax, which empowers radiologists to interact with images using natural language prompts and questions, similar to how they converse with colleagues. Furthermore, our model is equipped with the knowledge of identifying pathologies and diagnostic reasoning. D-Rax leverages established AI models [11] to incorporate expert model diagnostic predictions for multiple diseases, thus reducing the risk of missed findings and aiding in achieving more accurate diagnoses. To exemplify the utility of our domain-specialist VLM, we chose chest X-ray (CXR) images for this study. CXRs are among the most commonly performed imaging studies and play a crucial role in the diagnosis and management of a wide range of medical conditions, including respiratory diseases, cardiac abnormalities, and thoracic injuries. The novelty and contributions of this work can be summarized as:

- *Enhanced instruction-following training with expert model predictions.* We introduce a novel, domain-specific, and multi-modal instruction-following training strategy enriched with multiple expert model predictions for large VLMs.

- *Expert-enhanced instruction-following data generation.* We use MIMIC-CXR and Medical-Diff-VQA datasets to generate baseline instruction-following data for the design of conversational image analysis tools. State-of-the-art (SOTA) AI (expert) models are incorporated to add diagnostic and demographic predictions to the baseline, thus creating expert-enhanced visual instruction-following training data.
- *D-Rax.* Our expert-enhanced instruction-following training leads to a more accurate radiologic assistance tool, demonstrated by comprehensive comparisons. The same training paradigm can potentially benefit other conversational AI tools.

2   Related Work

The introduction of foundational large VLMs has flooded the gates for the design of complex multi-modal AI tools. Flamingo [1] is one of the earliest multimodal VLMs that bridged the gap between image-only and text-only methods. It combines prompts and multi-line chains of thought to produce sensible outcomes. Another notable example is the Large Language and Vision Assistant (LLaVA) [20] model that leverages a multi-modal architecture capable of processing both visual and textual information. Both of these VLM frameworks closely follow the technicalities from the Contrastive Language-Image Pretraining (CLIP) [22] model, which is a technique to associate images with corresponding textual descriptions. Such VLMs are widely adopted in the computer vision industry and are a gateway to many advances in biomedicine.

In the realm of biomedical VLMs, BioMedClip [27] is an important foundation model, with vision-language processing capabilities, enabling several standard biomedical imaging tasks such as classification and visual questionanswering. LLaVA-Med [18], a specialized version of LLaVA, is tailored for biomedical applications, including radiology, to enable clinicians to interact with medical images in a conversational language setting, thereby facilitating more efficient radiological workflows. OphGLM [5] combined expert model deductions with large language models (LLM) by generating a diagnostic report from retinal images. Most biomedical VLMs, however, are generalized and suffer from hallucinations, inaccurate diagnosis, and imprecise question answering. A domainspecialized tool in radiology can help overcome these challenges and provide accurate outcomes.

3   Methods

3.1   Data

Baseline Instruction-following Data. The multi-modal nature of our task requires both vision and language information. In this study, we use the MIMIC-CXR and Medical-Diff-VQA datasets to generate a baseline instruction-following dataset for our experiments. MIMIC-CXR [7, 14, 15] is a large open-access dataset of 377, 110 CXRs with structured labels on cardiopulmonary conditions derived from 227, 827 free-text radiology reports. Medical-Diff-VQA [9, 10, 14] is a derivative of the MIMIC-CXR dataset containing 700, 703 question-answer (QA) pairs derived from CXRs. The questions are divided into seven categories: abnormality, presence, view, location, level, type, and difference. Each category can hold either open-ended questions such as 'why, what, how' etc. with dynamic natural language answers or close-ended questions such as 'Is there' with binary answers like 'yes/no'. To limit the complexity of the evaluation, we did not focus on longitudinal changes, therefore the difference QAs were removed from the current evaluation. As a result, only a single image per patient was extracted to form the test set. Table 1 summarizes the data distribution for the baseline dataset.

Table 1. Baseline instruction-following data - Summary of train and test datasets and percentage distribution of QA categories.

		Total	%Abnormality	Presence	View	Location	Level	Type
Train	#QA Pairs	429, 000	27.1	29.1	10.5	15.7	12.5	5.1
129, 232	#Open	219, 305	24.6	0	10.2	30.6	24.5	10.1
images	#Close	209, 695	29.8	59.4	10.8	0	0	0
Test	#QA Pairs	13, 688	26.8	29.3	13.5	14	11.6	4.8
4, 190	#Open	6, 683	23.6	0	14	28.8	23.7	9.9
images	#Close	7, 005	29.8	57.2	13	0	0	0

Enhanced Expert Instruction-Following Data. We enhanced the baseline dataset by incorporating MIMIC-CXR along with QA conversations and integrating expert model predictions using pre-trained models from the TorchXRayVision [3] model zoo. Expert predictions on the MIMIC-CXR dataset fall into one of the four categories - diseases, age, race, and view (Table 2). The outcomes of these SOTA AI model predictions are appended to the baseline dataset to create our expert-model enhanced instruction-following dataset. The medical conditions in the first category include cardiomegaly, atelectasis, pneumonia, infiltration, fracture, enlarged cardio mediastinum, lung opacity, pneumothorax, emphysema, hernia, lung lesion, pleural thickening, edema, effusion, fibrosis, nodule, mass, and consolidation.

Table 2. Expert-model enhanced instruction-following data - Details on the AI model, training dataset, and labels used for each category.

Expert Predictions	Model	Dataset	Labels
Disease diagnosis	DenseNet121 [11]	MIMIC-CXR	CheXpert[13]
Patient age	Regression [12]	NIH ChestX-ray8 [24]	
Patient race	Classifier [6]	MIMIC-CXR	Asian, Black, White
CXR view position	ChestViewSplit [26]		Frontal, Lateral

3.2   Domain Specific Radiologic Assistant Design

The original LLaVA-Med model was trained on 15 million figure-caption pairs from PubMed [27]. While this teaches the model the context of biomedical application, we argue that for the sensitive process of medical imaging diagnosis, it is beneficial to develop a domain-specific VLM. Therefore, we perform end-to-end instruction tuning by training our model with CXRs and VQA-derived instructions generated from the associated radiology reports. In the process, we generated novel and enhanced instruction-following data for CXRs by incorporating predictions from expert models (Fig. 1).

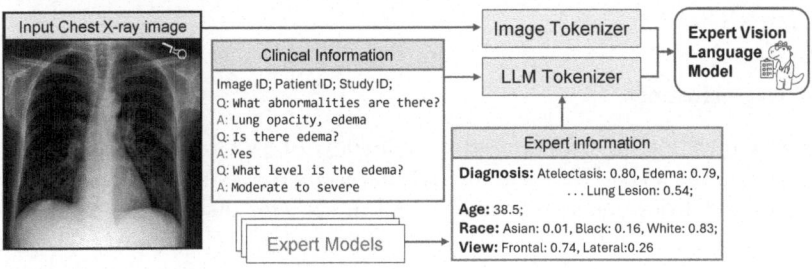

Fig. 1. Overview of our expert vision language model D-Rax design - Training data includes multimodal data including visual information (Chest X-ray images) and textual information (VQA from radiology reports, and expert model predictions).

Network Architecture: The definition of the expert VLM model follows the network architecture proposed in [20]. We chose Llama2 [23] as our LLM due to the availability of the pre-trained checkpoints and particularly used the Llama2-7B model. The visual encoder was kept consistent as ViT-Large/14 which is a pre-trained CLIP model. For any given input image X_v and a series of question and answer defined as $(X_q^1, X_a^1, X_q^2, X_a^2 ... X_q^T, X_a^T)$. First, X_v is transformed into a set of visual features Z_v by the CLIP model. For training with the instruction tuning data, the visual encoder is kept frozen and a trainable projection matrix W is used to convert the visual features Z_v into language embedding tokens $H_v = W.Z_v$ that can be jointly used with the language embedding tokens H_q of the questions X_q. The output of the model is X_a. Please note that H_q and H_v are used as inputs to the LLM model of the entire framework. For VLM training, we used the LLaVA-v1.5-7B [20] model as a baseline and the model weights were initialized with Llama2-7B weights added with a delta from LLaVA training.

Expert Model Enhanced Instruction Tuning Data: Within the Medical-Diff-VQA data, for any given input image X_v, there exists a multi-turn conversation that pertains to different categories of questions related to abnormality, presence, view, location, level, and type such that X_q belongs to a specific category. The dataset was enhanced with expert response X_e to the questions which when updated reads as $(\{X_e, X_q^1\} : X_a^1, \{X_e, X_q^2\} : X_a^2, ...\{X_e, X_q^T\} : X_a^T)$. Medical-Diff-VQA also provides QA pairs for difference (with a reference image), which we have not used in our experiments.

End-to-end Training: The complete training of domain-specific VLMs (particularly LLaVA) involves two steps: (1) concept alignment to biomedical concepts from large data from PubMed including figures in published articles, captions, and inline references to figures, and (2) an instruction tuning step, where both the projection layer and the language model are updated. In our method, we perform the instruction fine tuning with the multi-modal, expert-enhanced dataset for CXRs presented in Sect. 3.1. We also perform a set of experiments to establish the usefulness of employing expert model predictions to guide the radiology assistant's answers. Finally, we evaluate the performance of open- and close-ended conversational questions to establish the efficacy of our proposed strategy.

3.3   Experiments

For visual question answers related to radiology, LLaVA-Med was finetuned and evaluated on the VQA-RAD and SLAKE datasets. However, the data used covers multiple modalities and is relatively small in size, for instance: VAQ-RAD [16] has 315 radiology images and 3,515 QA pairs, and SLAKE [19] has 642 images and 7,000 QA pairs. For D-Rax, we utilize MIMIC, one of the largest domain-specific medical imaging data, and the associated VQA pairs. We further leverage expert models to provide more context and expert knowledge of the language model. We hypothesize that with expert model prediction, the trained radiologic assistant will have better outcomes in terms of reducing hallucinations and providing more precise and correct responses.

To establish the efficacy of utilizing model predictions related to abnormality, age, race, and view, we ran multiple experiments for end-to-end instruction tuning of various LLaVA models. In particular, we use the following pretrained models: LLaVA, LLaVA-Med finetuned on VQA-RAD (LLaVA-Med-RAD), and LLaVA-Med finetuned on SLAKE (LLaVA-Med-SLAKE). VQA-RAD and SLAKE are selected since they are closely related to our research question, however, the data represented has a much larger scope with fewer examples to enable the development of precise models. Overall we performed six different experiments, including end-to-end instruction fine-tuning with a model initialized with weights from the three aforementioned pre-trained models. For each of these model initializations, the instruction fine-tuning was performed both for the baseline dataset (images and VQA-derived instructions from MIMIC) and an enhanced dataset with augmentation of expert model predictions. The training was performed for a single epoch, with a learning rate of $2e^{-5}$ and an effective batch size of 8.

3.4   Evaluation

For performance evaluation, two metrics were utilized: accuracy and token recall, depending on the type of questions evaluated. For close-ended questions, the task can be considered as a classification, and hence we used accuracy. For open-ended questions, token recall measures the ratio of tokens correctly generated by the trained model according to the ground truth. Evaluating VLMs, particularly for open-ended questions is still a difficult problem and some approaches try to use OpenAI's GPT-4 to evaluate the similarity between ground truth and predicted answers [18, 20]. The inference of the finetuned model required 20G of GPU memory and could generate answers for 10, 000 questions per hour on a single NVIDIA H100 80G GPU.

4   Results

Performance of Enhanced Instruction. Figure 2 shows the qualitative evaluation of D-Rax by showing an example of conversations on a given CXR, as generated by VLMs trained on basic and expert-enhanced data. The results from quantitative evaluation (Table 3) indicates that the enhanced expert instruction training allows for statistically significant improvements in the model performance for abnormality and presence questions (both open and closed-ended). Meanwhile, for location, level, and type questions, where the expert model provides no explicit information, training on both basic and enhanced data mostly yields similar performance and even showcases improvements when using the LLaVA-Med-RAD model as the base. Intriguingly, in addressing the view questions, the expert model introduces different view information but does not affect the model's capacity to derive correct answers from images and questions. Overall, expert model-enhanced instruction training enables higher performance without impeding the pre-trained model's inherent ability to comprehend queries and images.

Comparison with Expert Models. D-Rax is not expected to outperform disease-specific expert models that are restricted to answering simple and close-ended (C) questions based on classification. Analysis of experiments on Abnormality (C) questions shows that the diagnostic accuracy of expert models of **70.4%** was comparable to VLMs (p-value > 0.08), except in 1/18 inferences where enhanced LLaVA-Med-RAD (Table 3) outperformed significantly the expert models (p-value $= 0.01$). However, identifying abnormalities is just one aspect of the VLM. VLMs can handle complex and nuanced questions, unlike expert models which cannot understand natural language queries.

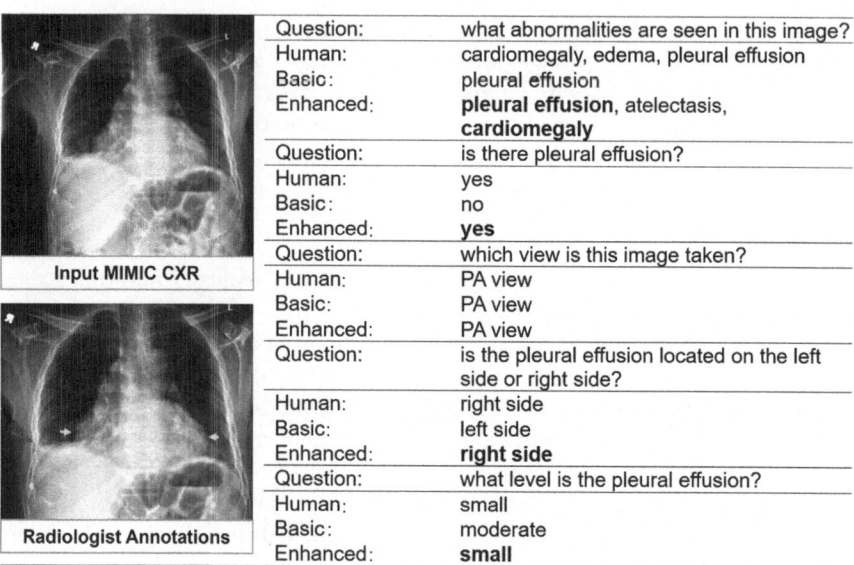

Question:	what abnormalities are seen in this image?
Human:	cardiomegaly, edema, pleural effusion
Basic:	pleural effusion
Enhanced:	**pleural effusion**, atelectasis, **cardiomegaly**
Question:	is there pleural effusion?
Human:	yes
Basic:	no
Enhanced:	**yes**
Question:	which view is this image taken?
Human:	PA view
Basic:	PA view
Enhanced:	PA view
Question:	is the pleural effusion located on the left side or right side?
Human:	right side
Basic:	left side
Enhanced:	**right side**
Question:	what level is the pleural effusion?
Human:	small
Basic:	moderate
Enhanced:	**small**

Fig. 2. Qualitative evaluation: conversations provided by VLMs trained on basic and expert enhanced data. The red arrow shows the area of the pleural effusion and the yellow arrows outline the lateral margins of the enlarged heart (cardiomegaly) provided by the radiologist, which were correctly identified by D-Rax.

Table 3. Quantitative evaluation: token recall (%) for open-ended questions (O) and accuracy (%) for close-ended questions (C) are reported to show the performance of domain-specific VLM with basic and enhanced instruction tuning strategies across various question types. Each value is an average and standard deviation of three inferences. The asterisks show statistical significance across paired comparisons using the Wilcoxon signed rank test (* for p-value < 0.05 and ** for p-value < 0.001).

Metrics (%)		LLaVA		LLaVA-Med-RAD		LLaVA-Med-SLAKE	
Question Type		Basic	Enhanced	Basic	Enhanced	Basic	Enhanced
Abnormality	(O)	40.6(0.5)	**41.7(0.3)**	39.8(0.5)	**41.7(0.1)***	39.5(0.5)	**42.0(0.7)****
	(C)	70.1(1.3)	**71.5(0.3)***	70.3(0.9)	**72.8(0.6)****	68.9(0.4)	**71.8(1.1)****
Presence	(C)	76.1(0.2)	**77.7(0.1)****	75.5(0.2)	**77.6(0.4)****	75.0(0.3)	**77.9(0.4)****
View	(O)	99.7(0.0)	99.7(0.0)	99.6(0.0)	99.6(0.0)	99.6(0.1)	99.6(0.1)
	(C)	99.0(0.2)	98.8(0.1)	98.9(0.2)	99.1(0.2)	98.8(0.1)	98.6(0.2)
Location	(O)	61.8(0.0)	61.6(0.4)	60.2(0.4)	**61.6(0.6)***	60.3(0.2)	**61.8(0.5)***
Level	(O)	59.1(0.8)	59.5(0.4)	58.8(0.5)	**60.4(0.4)***	59.2(0.8)	60.0(0.9)
Type	(O)	60.6(1.0)	60.6(0.8)	58.9(1.0)	58.5(1.0)	58.1(0.2)	58.4(1.3)
Average	(O)	61.3	61.6	60.4	**61.6****	60.4	**61.7****
	(C)	77.3	**78.6****	77.0	**79.0****	76.3	**78.8****

Ablation Studies. While we maintained the test set's characteristics by extracting one image per patient (Sect. 3.1), the following ablation study (Table 4) shows the results evaluated on an extended test set, including all the images from each patient. The improved results demonstrate the robustness of the method when tested on larger data. However, since evaluation of the larger test data is computationally expensive, most results reported in the paper are on the smaller test set.

Table 4. Evaluation on an extended test set. The asterisks show statistical significance across paired comparisons using the Wilcoxon signed rank test (* for p-value < 0.05 and ** for p-value < 0.001).

Question Type	Abnormality		Presence	View		Location	Level	Type	Average	
Metrics (%)	(O)	(C)	(C)	(O)	(C)	(O)	(O)	(O)	(O)	(C)
Test Set 4,190 images 13,688 QA pairs										
LLaVA-Basic	40.6	70.1	76.1	99.7	99.0	61.8	59.1	60.6	61.3	77.3
LLaVA-Enhanced	**41.7**	**71.5***	**77.7****	99.7	98.8	61.6	59.5	60.6	61.6	**78.6****
Extended Test Set 32,205 images 107,379 QA pairs										
LLaVA-Basic	42.7	73.4	76.1	99.5	98.7	61.8	57.7	60.0	59.9	77.7
LLaVA-Enhanced	**43.9***	**75.4****	**77.2****	99.5	98.7	61.9	**58.7***	59.9	**60.4***	**78.9****

5 Discussion and Conclusion

Our goal for developing D-Rax, a domain-specific expert model-guided radiologic assistant, is to reduce the hallucinations and improve the precision observed in responses from VLMs. We achieve this goal by establishing a novel training paradigm incorporating predictions from expert models. Hence, in our target application of CXR analysis, we embed expert predictions for disease, age, race, and view with the VQA instructions generated from radiological reports. Our results validate our hypothesis that (1) domain-specific knowledge, such as the use of MIMIC-CXR and Medical-Diff-VQA for CXR analysis, extracted from clinical radiology reports introduces a human factor into the model resulting in reduced hallucinations and allowing the system to provide precise information; and (2) addition of expert information from SOTA AI models generates statistically significant improved outcomes, enhancing accuracy of answering both open and close-ended questions in a conversation. D-Rax has the potential to enable a natural flow of diagnostic reasoning, enhance communication among clinicians, provide clear and accessible information to patients, and ultimately improve clinical care.

A Expert Enhanced Training

(See Fig. 3)

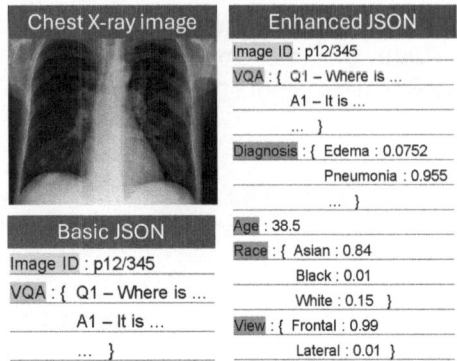

Fig. 3. Data organization for expert enhanced training containing the following information: (1) image identifiers, (2) question-answer pairs, (3) diagnostic prediction on 18 medical conditions, (4) predicted age of the patient, (5) predicted race of the patient, and (6) predicted view of the CXR.

B   No Abnormality Questions

(See Table 5)

Table 5. Removing abnormality questions (27% of the data) from training. Token recall (%) for open-ended questions (O) and accuracy (%) for close-ended questions (C) are reported to show the performance of LLaVA models finetuned on enhanced instruction dataset using 100% and 73% data, respectively. Each value is an average of three inferences and standard deviations are reported in parentheses. The asterisks show statistical significance across paired comparisons using the Wilcoxon signed rank test (* for p-value < 0.05 and ** for p-value < 0.001).

Metrics (%)		LLaVA-Enhanced		LLaVA-Med-RAD-Enhanced	
Question Type		100% data	73% data	100% data	73% data
Abnormality	(O)	**41.7(0.3)**	0.0(0.0)**	**41.7(0.1)**	0.0(0.0)**
	(C)	**71.5(0.3)**	43.1(0.7)**	**72.8(0.6)**	44.7(0.2)**
Presence	(C)	77.7(0.1)	**81.6(0.3)****	77.6(0.4)	**81.4(0.7)****
View	(O)	99.7(0.0)	99.8(0.0)	99.6(0.0)	99.7(0.1)*
	(C)	98.8(0.1)	**99.3(0.2)***	99.1(0.2)	98.9(0.2)
Location	(O)	61.6(0.4)	**64.5(0.4)****	61.6(0.6)	**64.5(0.3)****
Level	(O)	59.5(0.4)	59.5(0.7)	60.4(0.4)	59.9(1.0)
Type	(O)	60.6(0.8)	58.8(0.7)	58.5(1.0)	59.1(1.1)

C   Expert Model Metrics

(See Table 6)

Table 6. Quantitative evaluation of the expert model for disease diagnosis (DenseNet121) on 20% of MIMIC-CXR. The AUC performance is reported. https://github.com/mlmed/torchxrayvision/blob/master/BENCHMARKS.md

Atelectasis	Cardiomegaly	Consolidation	Edema	Enlarged Cardiomediastinum	Fracture	Lung Lesion	Lung Opacity	Effusion	Pneumonia	Pneumothorax
0.88	0.88	0.91	0.92	0.84	0.74	0.82	0.86	0.92	0.82	0.81

References

1. Alayrac, J.B., et al.: Flamingo: a visual language model for few-shot learning. Adv. Neural. Inf. Process. Syst. **35**, 23716–23736 (2022)
2. Bruls, R.J., Kwee, R.M.: Workload for radiologists during on-call hours: dramatic increase in the past 15 years. Insights Imaging **11**, 1–7 (2020)
3. Cohen, J.P., et al.: TorchXRayVision: a library of chest X-ray datasets and models. In: Medical Imaging with Deep Learning (2022)
4. Fawzy, N.A., et al.: Incidence and factors associated with burnout in radiologists: a systematic review. Eur. J. Radiol. Open **11**, 100530 (2023)
5. Gao, W., et al.: Ophglm: training an ophthalmology large language-and-vision assistant based on instructions and dialogue (2023). https://arxiv.org/abs/2306.12174
6. Gichoya, J.W., et al.: Ai recognition of patient race in medical imaging: a modelling study. The Lancet Digital Health (2022)
7. Goldberger, A., et al.: PhysioBank, PhysioToolkit, and PhysioNet: components of a new research resource for complex physiologic signals. Circulation **101**(23), e215–e220 (2000)
8. Hemmer, P., Schemmer, M., Riefle, L., Rosellen, N., Vössing, M., Kühl, N.: Factors that influence the adoption of human-AI collaboration in clinical decision-making. In: Thirtieth European Conference on Information Systems (ECIS 2022) (2022)
9. Hu, X., et al.: Medical-Diff-VQA: a large-scale medical dataset for difference visual question answering on chest x-ray images. PhysioNet (2023)
10. Hu, X., et al.: Expert knowledge-aware image difference graph representation learning for difference-aware medical visual question answering. In: Proceedings of the 29th ACM SIGKDD Conference on Knowledge Discovery and Data Mining, pp. 4156–4165 (2023)
11. Huang, G., Liu, Z., van der Maaten, L., Weinberger, K.Q.: Densely connected convolutional networks. In: Proceedings of the IEEE Conference on Computer Vision and Pattern Recognition (CVPR) pp. 2261–2269 (2017)
12. Ieki, H.E.A.: Deep learning-based age estimation from chest X-rays indicates cardiovascular prognosis. Commun. Med. (2022)
13. Irvin, J., et al.: Chexpert: A large chest radiograph dataset with uncertainty labels and expert comparison. In: Proceedings of the AAAI Conference on Artificial Intelligence, pp. 590–597 (2019)

14. Johnson, A., et al.: MIMIC-CXR-JPG - chest radiographs with structured labels. PhysioNet (2019)
15. Johnson, A.E.W., et al.: MIMIC-CXR-JPG, a large publicly available database of labeled chest radiographs. PysioNet (2019)
16. Lau, J.J., Gayen, S., Ben Abacha, A., Demner-Fushman, D.: A dataset of clinically generated visual questions and answers about radiology images. Sci. Data **5**(1), 1–10 (2018)
17. Lee, C.S., Nagy, P.G., Weaver, S.J., Newman-Toker, D.E.: Cognitive and system factors contributing to diagnostic errors in radiology. Am. J. Roentgenol. **201**(3), 611–617 (2013)
18. Li, C., et al.: LLaVA-med: training a large language-and-vision assistant for biomedicine in one day (2023)
19. Liu, B., Zhan, L.M., Xu, L., Ma, L., Yang, Y., Wu, X.M.: Slake: a semantically-labeled knowledge-enhanced dataset for medical visual question answering. In: 2021 IEEE 18th International Symposium on Biomedical Imaging (ISBI), pp. 1650–1654. IEEE (2021)
20. Liu, H., Li, C., Wu, Q., Lee, Y.J.: Visual instruction tuning (2023)
21. Mukherjee, P., Hou, B., Lanfredi, R.B., Summers, R.M.: Feasibility of using the privacy-preserving large language model vicuna for labeling radiology reports. Radiology **309** (2023)
22. Radford, A., et al.: Learning transferable visual models from natural language supervision (2021)
23. Touvron, H., et al.: Llama: open and efficient foundation language models (2023)
24. Wang, X., Peng, Y., Lu, L., Lu, Z., Bagheri, M., Summers, R.M.: Chestx-ray8: hospital-scale chest x-ray database and benchmarks on weakly-supervised classification and localization of common thorax diseases. In: Proceedings of the IEEE Conference on Computer Vision and Pattern Recognition, pp. 2097–2106 (2017)
25. Wu, C., Zhang, X., Zhang, Y., Wang, Y., Xie, W.: Towards generalist foundation model for radiology by leveraging web-scale 2d&3d medical data (2023)
26. Yi, X.: chestviewsplit. https://github.com/xinario/chestViewSplit
27. Zhang, S., et al.: Large-scale domain-specific pretraining for biomedical vision-language processing (2023)

Optimal Prompting in SAM for Few-Shot and Weakly Supervised Medical Image Segmentation

Lara Siblini[1,2], Gustavo Andrade-Miranda[1(✉)] [ID], Kamilia Taguelmimt[1],
Dimitris Visvikis[1], and Julien Bert[1]

[1] LaTIM, UMR1101, INSERM, University of Brest, Brest, France
`andradema@univ-brest.fr`
[2] Université de Bourgogne, Dijon, France

Abstract. Recent advancements in medical image segmentation have been driven by deep learning's capability to extract rich features from extensive datasets. However, these improvements rely heavily on large annotated datasets, which pose significant challenges in the resource-intensive medical field. Foundational models, such as Meta's Segment Anything Model (SAM), have been developed to address these challenges. SAM has demonstrated exceptional zero-shot performance, often rivaling or surpassing fully supervised models across various tasks. Nonetheless, SAM cannot be directly applied to medical image segmentation due to domain shift, making it necessary to fine-tune the model using prompts. Reducing the annotation workload is crucial to alleviate the burden and constraints associated with extensive data annotation in the medical field. This study investigates prompt-guided strategies in SAM for medical image segmentation under few-shot and weakly supervised scenarios. We assess various strategies-bounding boxes, positive points, negative points, and their combinations-using two publicly available datasets. Optimal results are achieved using positive-negative points, demonstrating that the SAM model can perform comparably to established methods in hepatic vascular and prostate cancer segmentation, even with minimal examples. This research aims to advance medical image segmentation by decreasing reliance on extensive annotated data, providing insights into effective prompt utilization, and showcasing SAM's adaptability in specialized medical contexts.

Keywords: Foundational models · prompt tuning · SAM · segmentation

1 Introduction

Recent advancements in medical image segmentation have been driven by significant progress in deep learning techniques [3]. The remarkable potential of deep learning lies in its ability to automatically learn features from extensive

---

L. Siblini and G. Andrade-Miranda-Authors contributed equally.

© The Author(s), under exclusive license to Springer Nature Switzerland AG 2025
Z. Deng et al. (Eds.): MedAGI 2024, LNCS 15184, pp. 103–112, 2025.
https://doi.org/10.1007/978-3-031-73471-7_11

datasets, leading to substantial improvements in performance [1]. However, these advancements come at the cost of requiring large quantities of annotated data to achieve optimal results [15]. This dependency on large, annotated datasets poses challenges, particularly in the medical field where acquiring and annotating high-quality data is time-consuming and resource-intensive [20]. Furthermore, conventional segmentation models are often trained within task-specific frameworks, which may not always have access to extensive datasets. This scarcity of data is reinforced by a lack of inter-center variability, restricting the models' ability to generalize to unseen data from different institutions [18]. As a result, these models frequently need to be retrained from scratch for each new application or dataset, further increasing the burden on resources and time.

To overcome the limitations of traditional training methods, many studies have employed transfer learning from pre-trained data on ImageNet [13]. However, this approach often resulted in suboptimal performance when applied to the medical domain [16]. In response, self-supervised learning [2] has emerged as a new paradigm enabling models to learn accurate and meaningful representations of input data without labels, thereby facilitating the development of large-scale foundational models. These comprehensive models, trained on extensive corpora, represent a significant breakthrough in the AI community. They can be easily adapted to task-specific problems through fine-tuning and prompting strategies [4,6], significantly reducing the reliance on extensive domain-specific training data. One of the most notable examples of these models is the Segment Anything Model (SAM) from Meta [12], which was trained on an extensive corpus comprising 1 billion annotations across 11 million images.

SAM can segment objects using various human input prompts like dots, bounding boxes, or text. Evaluations highlight its exceptional zero-shot performance, often matching or surpassing fully supervised models across diverse tasks [19]. With these strengths, SAM holds promise for various vision applications, including medical image segmentation. However, due to the substantial domain shift between natural and medical images, directly applying SAM is impractical [8]. Therefore, fine-tuning the SAM model with optimal prompting strategies is crucial for achieving accurate region segmentation. Previous studies have evaluated the zero-shot performance of SAM [14]. However, our focus is on fine-tuning SAM using few-shot learning and optimized prompting strategies. This approach aims to alleviate the burden of extensive manual labeling by leveraging weakly supervised segmentation with minimal annotated data through points or box prompts. This method enhances the efficiency of annotating large datasets and improves segmentation accuracy.

In this study, we explore prompt-guided strategies within the SAM model for medical image segmentation under few-shot and weakly supervised learning scenarios. Our primary contributions can be summarized as follows:

– We conduct an extensive assessment to evaluate various prompting strategies in the context of few-shot learning, with the aim of enhancing medical segmentation performance for hepatic vascular and prostate cancer.

- We compare the performance of the best prompting strategy with that of fully supervised state-of-the-art methods trained on limited data. This comparison involves two transformer-based segmentation networks and the well-known nnU-Net framework [11].
- We provide insights into the optimal location, position, and number of points required to enhance segmentation performance. Additionally, we discuss the potential benefits and challenges associated with each prompting method, offering a comprehensive evaluation of their efficacy in hepatic vascular and prostate cancer segmentation.

2 Methods

2.1 Overview of SAM

SAM represents the largest foundational model for natural image segmentation, trained on the extensive SA-1B dataset. It is designed to handle various user prompts for image segmentation. SAM consists of three core components: an Image Encoder (IE) based on the Vision Transformer (ViT) architecture [5], which extracts features from images; a Prompt Encoder (PE) that processes different types of prompts, including points, bounding boxes, masks, and text; and a lightweight Decoder (MD) that translates the combined image and prompt embeddings into segmentation results.

2.2 Prompting Strategies

In the following subsections, we will explore various prompting strategies designed to guide the foundational SAM model in medical image segmentation. These strategies include the use of bounding boxes, positive points, simultaneous positive and negative points, and a hybrid approach that integrates both bounding boxes and points. Each strategy provides different levels of guidance to the model, aiming to enhance segmentation accuracy while reducing reliance on extensive annotated datasets.

Box Prompting. Box prompting uses bounding boxes around regions of interest to train the model, providing a weakly supervised approach to guide the segmentation process. These bounding box coordinates are generated based on the ground truth segmentation. To emulate annotator variability, the bounding boxes are randomly expanded by 5 to 20 pixels around the ground truth. This expansion extends the bounding box slightly beyond the non-zero elements, introducing a form of human variability in the bounding box dimensions.

Point Prompting. This strategy involves using labeled points to guide the model in the segmentation process. Each point is associated with a label: positive points, situated inside the region of interest (ROI), are labeled as +1 (depicted

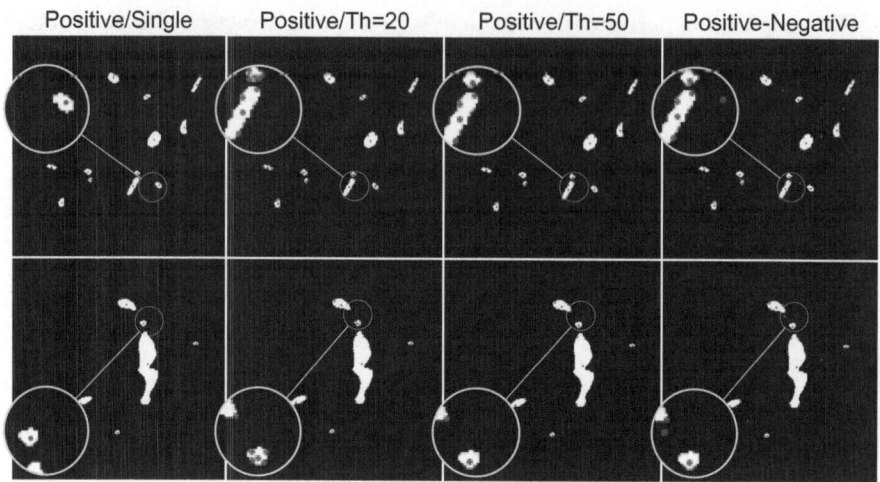

Fig. 1. Illustration of different point prompting tuning strategies. The first column represents single positive points. The second and third columns depict multiple positive points with different Th. The last column represents the positive-negative strategy with a threshold of $Th = 50$. Positive points are represented in red, while negative points are shown in green.

as red dots in Fig. 1), while negative points, located outside the ROI and corresponding to the background, are labeled as –1 (depicted as green dots in Fig. 1). Generally, we can differentiate between two strategies: utilizing solely positive points or integrating both positive and negative points.

Positive points prompting: This method involves placing one or several points within the target regions. In our experiments, we tested both strategies. In the first strategy, each disjoint region is represented by a single point positioned near the centroid of the object. To reflect annotator variability, the points are placed randomly within a circle of radius r, where r is smaller than the radius of the largest inscribed circle. The first column of Fig. 1 illustrates two examples of single-point prompting, with each positive point randomly positioned near the object's centroid to simulate realistic annotation conditions.

To determine the use of single or multiple points per region, we base our decision on the area of each object to be segmented. We evaluate a series of thresholds Th and decide whether to include additional points based on these values. When an object's area exceeds Th, additional points are randomly distributed along the contours with a margin of error relative to the ground truth. This method enhances segmentation in larger regions and those with varying contrast, thereby avoiding under-segmentation problems. The second and third columns in Fig. 1 illustrate this effect for threshold values of $Th = 20$ and $Th = 50$. The smaller region in the second column contains more points compared to the same region in the third column, where $Th = 50$. However, in both cases, the larger region consistently contains more than one point.

Positive-negative points prompting: This method works similarly to the previous one, with the distinction of also including negative points. Positive points are assigned based on the same threshold principle, while negative points are strategically placed between the contours of nearly disjoint regions, where the model is prone to misinterpreting them as a single region. The objective is to mitigate over-segmentation issues that arise due to the presence of closely situated objects. The last column of Fig. 1 illustrates this concept, showing a negative point (green) placed between two closely situated objects.

Hybrid - Box-Points Prompting. In the final implementation, a combination of bounding boxes and positive-negative points was used. This strategy incorporates the previously described method of assigning multiple positive and negative points along with bounding boxes. This hybrid method aims to leverage the strengths of both point-based and bounding box techniques to achieve more precise and reliable segmentation results.

3 Experiments

3.1 Datasets

In our experiments, we employ the 3D-IRCADb [7] and PICAI [17] datasets to evaluate the SAM model. For both datasets, we perform evaluations in the context of few-shot learning, focusing on scenarios with a limited number of annotated samples.

3D-IRCADb: This dataset comprises 20 contrast-enhanced CT volumes with varying image resolutions and vessel structures, featuring hepatic tumors in 75% of the cases. For this study, images were resized axially using the liver bounding box in each CT scan to a resolution of 256 × 256 with 128 slices per patient. The original anisotropic voxel spacing was standardized to an isotropic spacing of $1 \times 1 \times 1$ mm.

PICAI: This dataset, obtained from the corresponding grand challenge, comprises over 1500 publicly available cases. Each case features three imaging modalities: T2-weighted imaging (T2W), diffusion-weighted imaging (DWI), and apparent diffusion coefficient maps (ADC). Specifically, we examine 40 cases where clinically significant prostate cancer (csPCa) is identified.

3.2 Implementation Details

We experimented with various prompting strategies, including bounding boxes, single and multiple positive points, positive-negative points, and combinations of boxes and points. For all configurations, we fine-tuned only the mask decoder, keeping the image and prompt encoder frozen. All experiments were conducted using the ViT-B model for inference and fine-tuning, treating the final segmentation task as binary for both datasets. Models were fine-tuned for 200 epochs using the Adam optimizer with an initial learning rate of 1e–5, and other hyperparameters set according to the original SAM paper. Training was conducted with a batch size of 10 on a single GPU with 48GB of memory. For the 3D-IRCADb dataset, we trained and validated on 18 volumes and tested on 2 volumes. For the PICAI dataset, we used 20 volumes for training and validation, and an additional 20 for testing.

3.3   Evaluation

We first compare various prompting strategies with three state-of-the-art (SOTA) models: Swin-UNETR [9], UNETR [10], and nnU-Net [11]. This evaluation is conducted on the 3D-IRCADb dataset, assessing Dice, Jaccard, and Precision metrics. All models are trained using the same number of volumes. After identifying the optimal prompting strategy, we evaluate its generalizability by applying it to a new downstream task: prostate tumor segmentation using the PICAI dataset. For this assessment, we employ the Dice score as the evaluation metric. To further evaluate the few-shot capabilities of the optimal prompting strategy, we fine-tuned the model on varying numbers of training cases (ranging from 6 to 16). Additionally, we present qualitative results for both the PICAI and 3D-IRCADb datasets, visually comparing the performance of the prompting strategies and state-of-the-art methods. This comparison aims to highlight the strengths of the prompting approach and demonstrate the effectiveness of the optimal strategy across different scenarios.

4   Results and Discussion

Initially, we evaluated the performance of various prompting strategies on the 3D-IRCADb dataset using two test volumes (see Table 1). The best performance was achieved with Positive-Negative prompting, yielding Dice scores of 0.74 and 0.70. Close to this strategy were the hybrid and multiple positive points approaches, with Dice scores of 0.72 and 0.68, respectively. Among the state-of-the-art (SOTA) methods, the best-performing model was nnU-Net, achieving Dice scores of 0.74 and 0.60. It is important to note that all prompting strategies and supervised methods were trained using the same amount of data (18 samples). Particularly noteworthy is that the worst-performing prompting strategy was the one based on bounding boxes. This may be due to the inclusion of false positive pixels within the bounding box, which can negatively impact the segmentation task.

We also investigated whether the use of a positive-negative points prompting, maintains good performance in a different downstream task. We then compared this strategy against SOTA models, with the results presented in Fig. 2. It is evident that SAM significantly outperforms the SOTA models, with all predictions achieving Dice scores of at least 0.6. These results underscore the importance of foundational models and the effectiveness of prompt tuning, especially in data-limited scenarios.

Table 1. Results of the different prompting strategies and SOTA methods for the two test volumes in the 3D-IRCADb dataset.

Prompting	Dice (↑)		Jaccard (↑)		Precision (↑)	
	S1	S2	S1	S2	S1	S2
Box	0.54	0.48	0.40	0.33	0.71	0.61
Positive point/Single	0.67	0.62	0.52	0.46	0.76	0.69
Positive point/Th = 50	0.70	0.64	0.55	0.50	0.82	0.71
Positive point/Th = 20	0.72	0.67	0.57	0.53	0.83	0.71
Positive-Negative/Th=20	**0.74**	**0.70**	**0.60**	**0.55**	0.83	0.74
Hybrid	0.72	0.68	0.58	0.53	0.81	0.71
UNETR	0.51	0.43	0.34	0.28	0.56	0.65
nnU-Net	**0.74**	0.60	0.59	0.43	**0.90**	**0.81**
Swin-UNETR	0.63	0.51	0.46	0.34	0.76	0.74

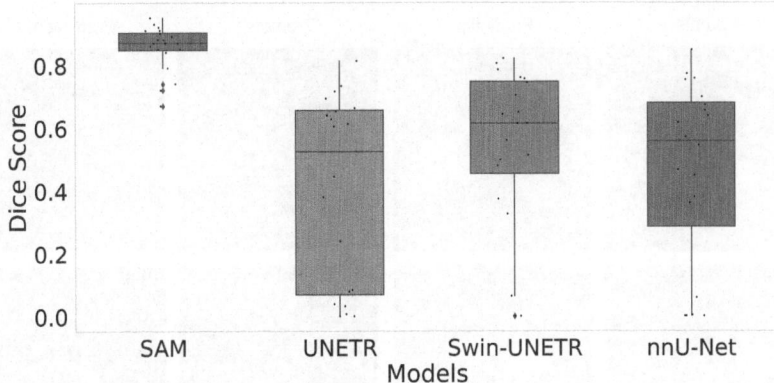

Fig. 2. Boxplots comparing the Positive-Negative prompting strategy with UNETR, Swin-UNETR, and nnU-Net models for PICAI datasets.

To evaluate the few-shot capabilities of the positive-negative prompting strategy, we trained SAM varying the numbers of training cases. The results in Fig. 3 show that the positive-negative fine-tuning outperformed supervised methods even with only six training cases. The only exception is in $S1$, where nnU-Net performs better. These findings highlight the robust few-shot learning capability of the SAM model. Performance improvements became significant after using more than 10 training cases. However, further investigation is needed to determine the plateau point beyond which additional training samples do not yield substantial gains.

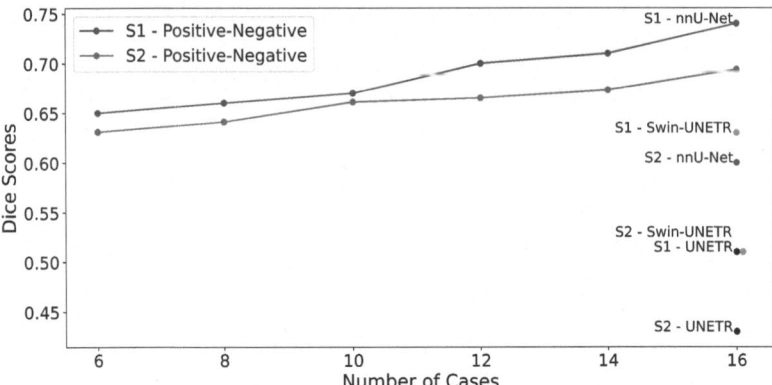

Fig. 3. Illustration of Dice scores on the `3D-IRCADb` dataset using the positive-negative prompting strategy. Lines represent performance with varying training sample sizes (6 to 16). Colored dots indicate Dice scores of SOTA methods for comparison.

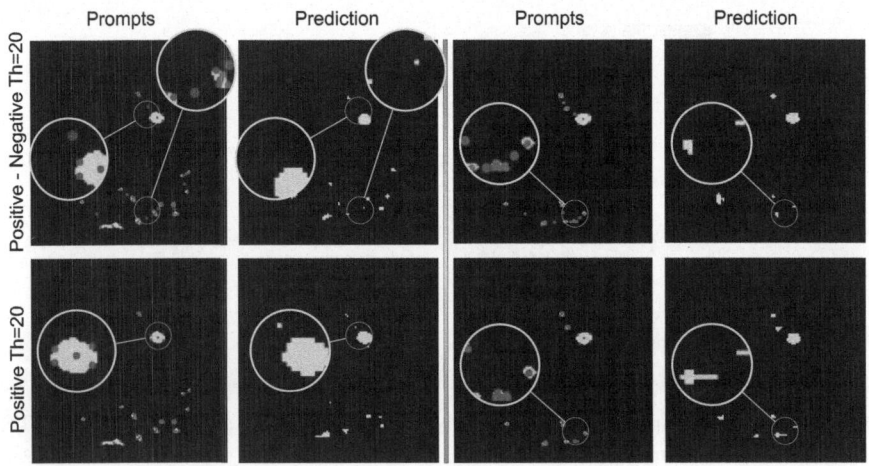

Fig. 4. Visual comparison of positive-negative versus positive points prompting. Magnified areas are highlighted in yellow, with positive points shown in red and negative points in green. (Color figure online)

Figure 4 illustrates the difference between using only positive points and using both positive and negative points. For instance, the inclusion of negative points helps to accurately predict small regions that disappear when only positive points are used (first and second columns). Additionally, the absence of negative points (the last two columns) leads to small regions being segmented as a single entity, resulting in over-segmentation. In the prostate cancer segmentation task depicted in Fig. 5, although the models sometimes struggle to accurately delineate the shape of the tumor, they consistently detect its presence.

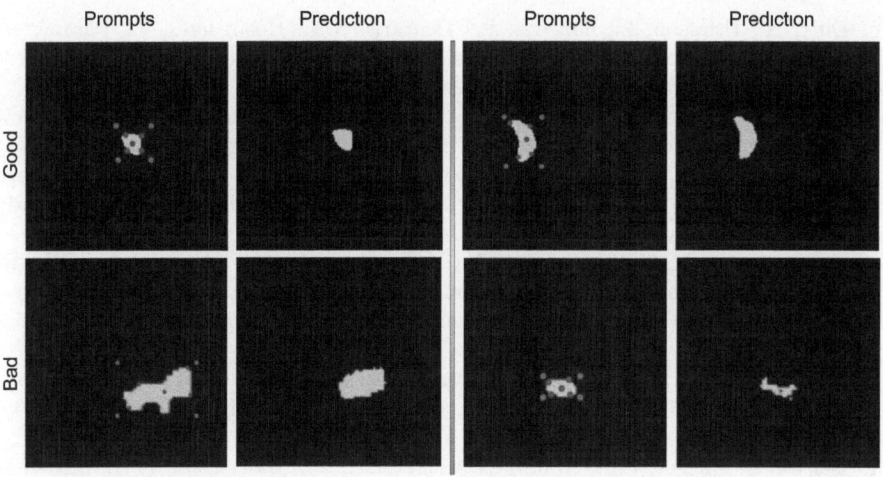

Fig. 5. Thumbnails of good and poor Segmentation results using positive-negative prompts on the `PICAI` Dataset

5   Conclusion

By employing various prompting fine-tuning strategies on the SAM model, we have demonstrated its versatility across two medical image segmentation tasks. Our study reveals that incorporating both positive and negative points is the most effective approach for enhancing segmentation performance. Specifically, our findings indicate that multiple points may be necessary depending on the size of the object, particularly near its boundary. The inclusion of negative points also helps to mitigate over-segmentation and under-segmentation issues, especially when disjoint objects are in close proximity. This work provides valuable insights and guidelines for the optimal placement of prompts in the context of medical imaging, thereby reducing the reliance on fully annotated datasets.

References

1. Andrade-Miranda, G., Jaouen, V., Tankyevych, O., Cheze Le Rest, C., Visvikis, D., Conze, P.H.: Multi-modal medical transformers: ameta-analysis for medical image segmentation in oncology. Comput. Med. Imaging Graph. **110**, 102308 (2023)
2. Balestriero, R., et al.: A cookbook of self-supervised learning (2023)
3. Conze, P.H., Andrade-Miranda, G., Singh, V.K., Jaouen, V., Visvikis, D.: Current and emerging trends in medical image segmentation with deep learning. IEEE Trans. Radiat. Plasma Med. Sci. **7**(6), 545–569 (2023)
4. Davila, A., Colan, J., Hasegawa, Y.: Comparison of fine-tuning strategies for transfer learning in medical image classification. Image Vis. Comput. **146**, 105012 (2024)
5. Dosovitskiy, A., et al.: An image is worth 16 x 16 words: transformers for image recognition at scale. In: International Conference on Learning Representations (ICLR) (2021)

6. Dutt, R., Ericsson, L., Sanchez, P., Tsaftaris, S.A., Hospedales, T.: Parameter-efficient fine-tuning for medical image analysis: The missed opportunity. In: Medical Imaging with Deep Learning (2024)
7. Garret, G., Vacavant, A., Frindel, C.: Deep vessel segmentation based on a new combination of vesselness filters. ArXiv **abs/2402.14509** (2024)
8. Gu, H., Dong, H., Yang, J., Mazurowski, M.A.: How to build the best medical image segmentation algorithm using foundation models: a comprehensive empirical study with segment anything model (2024)
9. Hatamizadeh, A., Nath, V., Tang, Y., Yang, D., Roth, H., Xu, D.: Swin UNETR: Swin Transformers for semantic segmentation of brain tumors in MRI images. In: Brainlesion: Glioma, Multiple Sclerosis, Stroke and Traumatic Brain Injuries (2022)
10. Hatamizadeh, A., et al.: UNETR: Transformers for 3D medical image segmentation. In: IEEE/CVF Winter Conference on Applications of Computer Vision, pp. 272–284 (2022)
11. Isensee, F., Jaeger, P.F., Kohl, S.A.A., Petersen, J., Maier-Hein, K.H.: nnU-Net: a self-configuring method for deep learning-based biomedical image segmentation. Nat. Methods **18**(2), 203–211 (2020)
12. Kirillov, A., et al.: Segment anything. arXiv:2304.02643 (2023)
13. Krizhevsky, A., Sutskever, I., Hinton, G.E.: Imagenet classification with deep convolutional neural networks. In: Pereira, F., Burges, C.J.C., Bottou, L., Weinberger, K.Q. (eds.) Advances in Neural Information Processing Systems, vol. 25. Curran Associates, Inc. (2012)
14. Mazurowski, M.A., Dong, H., Gu, H., Yang, J., Konz, N., Zhang, Y.: Segment anything model for medical image analysis: an experimental study. Med. Image Anal. **89**, 102918 (2023)
15. Moor, M., Banerjee, O., Abad, Z.S.H., et al.: Foundation models for generalist medical artificial intelligence. Nature **616**, 259–265 (2023)
16. Raghu, M., Zhang, C., Kleinberg, J., Bengio, S.: Transfusion: understanding transfer learning for medical imaging. Adv. Neural Inform. Process. Syst. **32** (2019)
17. Saha, A., et al., henkjan huisman: Artificial intelligence and radiologists at prostate cancer detection in MRI — the PI-CAI challenge. In: Medical Imaging with Deep Learning, short paper track (2023)
18. Sallé, G., et al.: Cross-modal tumor segmentation using generative blending augmentation and self-training. IEEE Trans. Biomed. Eng. 1–12 (2024)
19. Wald, T., et al.: Sam.md: Zero-shot medical image segmentation capabilities of the segment anything model. In: Medical Imaging with Deep Learning, short paper track (2023)
20. Yang, F., et al.: Assessing inter-annotator agreement for medical image segmentation. IEEE Access **11**, 21300–21312 (2023). epub 2023 Feb 27

UniCrossAdapter: Multimodal Adaptation of CLIP for Radiology Report Generation

Yaxiong Chen[1,2], Chuang Du[1], Chunlei Li[3], Jingliang Hu[3], Yilei Shi[3], Shengwu Xiong[1,2], Xiao Xiang Zhu[4], and Lichao Mou[3(✉)]

[1] Wuhan University of Technology,Wuhan, China
[2] Shanghai Artificial Intelligence Laboratory,Shanghai, China
[3] MedAI Technology (Wuxi) Co. Ltd.,Wuxi, China
lichao.mou@medimagingai.com
[4] Technical University of Munich,Munich, Germany

Abstract. Automated radiology report generation aims to expedite the tedious and error-prone reporting process for radiologists. While recent works have made progress, learning to align medical images and textual findings remains challenging due to the relative scarcity of labeled medical data. For example, datasets for this task are much smaller than those used for image captioning in computer vision. In this work, we propose to transfer representations from CLIP, a large-scale pre-trained vision-language model, to better capture cross-modal semantics between images and texts. However, directly applying CLIP is suboptimal due to the domain gap between natural images and radiology. To enable efficient adaptation, we introduce UniCrossAdapter, lightweight adapter modules that are incorporated into CLIP and fine-tuned on the target task while keeping base parameters fixed. The adapters are distributed across modalities and their interaction to enhance vision-language alignment. Experiments on two public datasets demonstrate the effectiveness of our approach, advancing state-of-the-art in radiology report generation. The proposed transfer learning framework provides a means of harnessing semantic knowledge from large-scale pre-trained models to tackle data-scarce medical vision-language tasks. Code is available at https://github.com/chauncey-tow/MRG-CLIP.

Keywords: report generation · CLIP · adapter

1 Introduction

Radiology report writing is a tedious and error-prone task for radiologists due to the large volume of images needing interpretation. Automated report generation has recently emerged as a promising solution to expedite this process and alleviate the workload for radiologists. This task bears similarity to image

---

C. Du—Work done during an internship at MedAI Technology (Wuxi) Co. Ltd.

© The Author(s), under exclusive license to Springer Nature Switzerland AG 2025
Z. Deng et al. (Eds.): MedAGI 2024, LNCS 15184, pp. 113–123, 2025.
https://doi.org/10.1007/978-3-031-73471-7_12

captioning in computer vision, whereby textual descriptions must be produced to characterize visual inputs.

There has been growing interest in this direction. The authors of [1] propose to generate radiology reports with a memory-driven Transformer and firstly conduct studies on MIMIC-CXR dataset [2]. They later augment their model with a cross-modal memory module [3,4] puts forth an approach to distill both posterior and prior knowledge to further boost performance. In order to better align visual and textual features, [5] employs reinforcement learning over the cross-modal memory network [3]. In [6], the authors design a cross-modal prototype network to facilitate interactions across modalities. Aiming to promote semantic alignment, [7] explicitly leverage text embeddings to guide visual feature learning. Recently, [8] introduces a framework that makes use of a dynamic graph to enhance visual representations in a contrastive learning paradigm for radiology report generation tasks.

Due to medical privacy concerns, the difficulty of gathering medical data, and the labor-intensive nature of annotation, the amount of data available for radiology report generation is relatively small compared to that used for image captioning in computer vision. For example, IU-Xray (4K images) [9] and MIMIC-CXR (220K images) [2] are much smaller than image captioning datasets Conceptual Captions (3.3M images) [10] and Conceptual 12M (12M images) [11]. Learning comprehensively from such limited data makes it challenging for current methods to fully understand cross-modal semantics between radiological images and reports [1,3–8]. Overcoming this paucity of labeled data to better learn these semantics is crucial for advancing radiology report generation.

Recently, leveraging large-scale pre-trained vision-language models, such as CLIP [12], which is trained on 400 million image-text pairs collected from the internet to match images with their corresponding textual descriptions, has become a promising approach for tackling downstream tasks in computer vision. However, the application of such models on radiology report generation still remains unexplored. In this work, we propose transferring the knowledge encapsulated in CLIP to the task of automatic report generation to better model the semantic relationship between medical images and their associated radiological findings.

Despite its strong performance, directly applying CLIP to radiology report generation tasks poses certain challenges. CLIP has been pre-trained on large-scale natural image-text datasets, exhibiting a substantial domain divergence from medical images. Therefore, while the model encapsulates rich semantic knowledge about everyday scenes, fine-tuning is imperative to adapt CLIP to radiology. However, conducting a full fine-tuning of a model as massive as CLIP is highly impractical given immense computational demands. To enable efficient adaptation, we propose uni- and cross-modal adapter (UniCrossAdapter), a parameter-efficient fine-tuning approach to adapt CLIP for the task of radiology report generation. The key idea is to integrate lightweight adapter modules into CLIP that can be fine-tuned on the target task while keeping the pre-trained backbone parameters frozen. The modules are distributed to both visual and

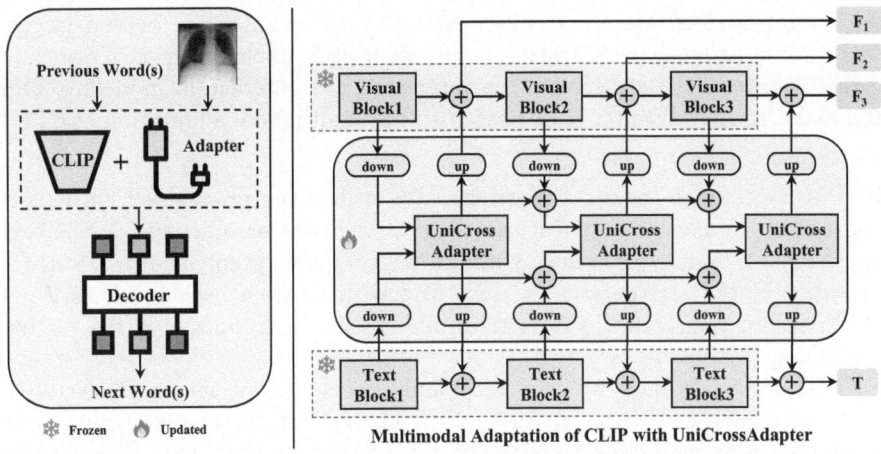

Fig. 1. (Left) Overall architecture of our method for radiology report generation, leveraging CLIP and the proposed UniCrossAdapter. (Right) Illustration of the interaction between the UniCrossAdapter and CLIP's text and image encoders.

textual modalities and their interactions for better aligning medical images and texts. Our contributions are three-fold.

- We investigate the transfer of representations learned by CLIP to describe medical image findings.
- We introduce a novel adapter architecture that improves vision-language alignment on radiology images and reports by coupling image and text adapter modules through a cross-attention mechanism.
- Our approach achieves state-of-the-art performance on IU-Xray and MIMIC-CXR, the two most used benchmark datasets.

2   Method

We propose an end-to-end framework for automatic radiology report generation, as illustrated in Fig. 1. The model comprises two key components: (i) the adaptation of CLIP with UniCrossAdapter to learn visual and textual representations for radiology data, and (ii) a decoder that generates reports. In what follows, we first detail each of them. Then, we describe the training and inference procedures.

2.1   Multimodal Adaptation of CLIP with UniCrossAdapter

Recent work has explored parameter-efficient fine-tuning methods [13–19] for adapting large pre-trained models to downstream tasks. However, architectures used in prior efficient tuning techniques, e.g., down-up feedforward layers [18,19] and LoRA [17], may be too simple to effectively adapt complex multimodal models. Moreover, most existing approaches have focused largely on unimodal or basic classification tasks, with little exploration on more challenging multimodal setups requiring inter-modality interaction modeling. Our proposed UniCrossAdapter is dedicated to the multimodal adaptation of CLIP.

CLIP's Text and Image Encoders. We utilize the pre-trained CLIP text Transformer to extract text features. Due to its large parameter size, the text Transformer remains frozen during fine-tuning. We then evenly split it into three sequential blocks and denote the text feature map from each block as $T_i \in \mathbb{R}^{N \times D}$, where $i \in \{1, 2, 3\}$, N is the number of tokens, and D is the feature dimension.

For the visual branch, we use CLIP's image encoder, specifically ResNet-101, to extract multi-scale visual features F_i from the last three stages. Similar to the text encoder, we freeze the image encoder during fine-tuning to leverage rich semantics learned from pre-training.

Unimodal and Cross-Modal Adaptation. The visual and linguistic features are first projected to a lower-dimensional space. Residual connections are further formed between consecutive adapter layers to enrich unimodal representations. This process can be formulated as

$$
\begin{aligned}
\hat{F}_i &= \mathrm{down}(F_i) + \hat{F}_{i-1} \,, \\
\hat{T}_i &= \mathrm{down}(T_i) + \hat{T}_{i-1} \,,
\end{aligned}
\tag{1}
$$

where $\mathrm{down}(\cdot)$ indicates dimension reduction layers implemented by convolutional and linear layers for visual and textual features, respectively. To encourage interactions within each modality, we apply multi-head self-attention (MHSA) on both modalities:

$$
\begin{aligned}
F_i^{sa} &= \mathrm{MHSA}(\hat{F}_i) \,, \\
T_i^{sa} &= \mathrm{MHSA}(\hat{T}_i) \,.
\end{aligned}
\tag{2}
$$

For coupling the visual and linguistic adapter modules, we perform multi-head cross-attention (MHCA) across the adapted unimodal representations for establishing cross-modal interactions:

$$
\begin{aligned}
F_i^{ca} &= \mathrm{FFN}(\mathrm{MHCA}(Q = F_i^{sa}, K = \hat{T}_i, V = \hat{T}_i)) \,, \\
T_i^{ca} &= \mathrm{FFN}(\mathrm{MHCA}(Q = T_i^{sa}, K = \hat{F}_i, V = \hat{F}_i)) \,.
\end{aligned}
\tag{3}
$$

Then, we incorporate the interacted features into the original features:

$$
\begin{aligned}
\tilde{F}_i &= \mathrm{up}(F_i^{ca}) + F_i \,, \\
\tilde{T}_i &= \mathrm{up}(T_i^{ca}) + T_i \,,
\end{aligned}
\tag{4}
$$

where $\mathrm{up}(\cdot)$ denotes dimension recovery implemented by deconvolution and linear layers.

Feature Modulation and Multi-scale Fusion. Since radiology images contain multi-scale anatomical structures (e.g., lung and heart) that require model attention, we fuse the multi-scale visual features to obtain comprehensive representations. Before fusion, we modulate the visual features of different scales by interacting a global text feature $\boldsymbol{\tau}$, obtained via a projection layer in the text Transformer, with each $\tilde{\boldsymbol{F}}_i$ to highlight relevant regions:

$$
\begin{aligned}
\boldsymbol{M}_i &= \mathrm{MHCA}(Q = s(\tilde{\boldsymbol{F}}_i), K = \boldsymbol{\tau}, V = \boldsymbol{\tau}), \\
\boldsymbol{Z} &= \mathrm{Conv}_{1\times1} \circ \mathrm{Concat}(\boldsymbol{M}_1, \boldsymbol{M}_2, \boldsymbol{M}_3),
\end{aligned}
\tag{5}
$$

where s denotes a convolutional layer to project the multi-scale features to a unified scale. $\boldsymbol{M}_i$ represents the modulated visual features. $\circ$ is a composition function, and $\boldsymbol{Z} \in \mathbb{R}^{C \times H \times W}$ is the fused visual feature.

In addition, to incorporate spatial information into $\boldsymbol{Z}$, we concatenate it with spatial coordinates $\boldsymbol{P} \in \mathbb{R}^{2 \times H \times W}$ across the channel dimension. The resulting feature is then passed through a 3×3 convolutional layer to reduce the enlarged channel dimension. This porcess can be written as

$$
\boldsymbol{X} = \mathrm{Conv}_{3\times3} \circ \mathrm{Concat}(\boldsymbol{Z}, \boldsymbol{P}).
\tag{6}
$$

Finally, we send $\boldsymbol{X}$ into a vision Transformer [20] network such that $\boldsymbol{X}$ is transformed to a sequence of feature vectors $\{\boldsymbol{v}_1, \boldsymbol{v}_2, \ldots, \boldsymbol{v}_N\}$, where $\boldsymbol{v}_i \in \mathbb{R}^D$ for the following procedure.

2.2 Report Decoder

We adopt a standard Transformer decoder [21] to generate reports. The decoder takes as input the adapted, fused multimodal representations from the CLIP-driven image and text encoders, and generates tokens autoregressively.

2.3 Training and Inference

Training. Let $\boldsymbol{I}$ be an input radiology image, and its ground truth report is denoted as $\boldsymbol{R} = \{[\text{SOS}], \boldsymbol{w}_1, \boldsymbol{w}_2, \ldots, \boldsymbol{w}_L, [\text{EOS}]\}$, where $\boldsymbol{w}_i \in \mathcal{V}$ represents the i-th token and $\mathcal{V}$ is the vocabulary set. [SOS] and [EOS] are the appended start and end tokens, while L is the length of the sequence. At training time, we first feed $\boldsymbol{I}$ and $\{[\text{SOS}], \boldsymbol{w}_1, \boldsymbol{w}_2, \ldots, \boldsymbol{w}_L\}$ into the image and text encoders with our adapter to derive a multimodal representation. The Transformer decoder then takes the multimodal representation as input and $\{[\text{SOS}], \boldsymbol{w}_1, \boldsymbol{w}_2, \ldots, \boldsymbol{w}_L\}$ as query to generate a predicted token sequence $\{\boldsymbol{p}_1, \boldsymbol{p}_2, \ldots, \boldsymbol{p}_L, \boldsymbol{p}_{L+1}\}$. We optimize the model by minimizing the cross entropy loss between the predicted sequence and the corresponding ground truth sequence $\{\boldsymbol{w}_1, \boldsymbol{w}_2, \ldots, \boldsymbol{w}_L, [\text{EOS}]\}$:

$$
\mathcal{L}_{\mathrm{ce}} = -\frac{1}{L+1} \sum_{i=1}^{L+1} \boldsymbol{w}_i \log(\boldsymbol{p}_i).
\tag{7}
$$

Inference. During inference, our model generates texts in an autoregressive manner. Given a test image, the model is first provided an [SOS] token as a prompt to predict the first token. The predicted first token is then concatenated with the [SOS] token as a new prompt to predict the second token. This process continues iteratively, with the previously predicted token(s) and [SOS] token as a prompt to predict each subsequent token, until an [EOS] token is predicted indicating the end of generation. This autoregressive way allows the model to condition each token prediction on its previous predictions, yielding more coherent and fluent text.

Table 1. Comparison results on the IU-Xray and MIMIC-CXR datasets. ∗ denotes results replicated from official code. † indicates replicated results without pre-training on the datasets. **Bold** indicates the best results, and <u>underline</u> indicates the second best results.

Dataset	Method	BLEU-1	BLEU-2	BLEU-3	BLEU-4	ROUGE-L	METEOR
IU-Xray	R2Gen	0.470	0.304	0.219	0.165	0.371	0.187
	SentSAT+KG	0.441	0.291	0.203	0.147	0.367	–
	CMCL	0.473	0.305	0.217	0.162	0.378	0.186
	$\mathcal{M}^2$ TR. PROG.	0.486	0.317	0.232	0.173	0.390	0.192
	CMN	0.475	0.309	0.222	0.170	0.375	0.191
	PPKED	0.483	0.315	0.224	0.168	0.376	–
	CMM+RL	0.494	0.321	<u>0.235</u>	<u>0.181</u>	0.384	0.201
	XPRONET∗	0.491	<u>0.325</u>	0.228	0.169	0.387	<u>0.202</u>
	DCL	–	–	–	0.163	0.383	0.193
	M2KT	<u>0.497</u>	0.319	0.230	0.174	**0.399**	–
	VLCI†	0.324	0.211	0.151	0.115	0.379	0.166
	RAMT	0.482	0.310	0.221	0.165	0.377	0.195
	PromptMRG	0.401	–	–	0.098	0.281	0.160
	Ours	**0.509**	**0.349**	**0.257**	**0.195**	<u>0.395</u>	**0.210**
MIMIC-CXR	R2Gen	0.353	0.218	0.145	0.103	0.277	0.142
	CMCL	0.344	0.217	0.140	0.097	0.281	0.133
	$\mathcal{M}^2$ TR. PROG.	0.378	0.232	0.154	0.107	0.272	0.145
	CMN	0.353	0.218	0.148	0.106	0.278	0.142
	PPKED	0.360	0.224	0.149	0.106	0.284	0.149
	CMM+RL	0.381	0.232	0.155	0.109	<u>0.287</u>	0.151
	XPRONET	0.344	0.215	0.146	0.105	0.279	0.138
	DCL	–	–	–	0.109	0.284	0.150
	M2KT	<u>0.386</u>	**0.237**	<u>0.157</u>	0.111	0.274	–
	VLCI†	0.357	0.216	0.144	0.103	0.256	0.136
	RAMT	0.362	0.229	0.157	<u>0.113</u>	0.284	<u>0.153</u>
	PromptMRG	**0.398**	–	–	0.112	0.268	**0.157**
	Ours	0.375	**0.237**	**0.165**	**0.120**	**0.289**	0.134

3   Experiments

3.1   Datasets and Evaluation Metrics

We conduct experiments on two datasets: IU-Xray [9] and MIMIC-CXR [2]. IU-Xray comprises 7,470 chest X-ray images along with 3,955 radiology reports. We tokenize words with > 3 occurrences and truncate/pad reports to 60 tokens. MIMIC-CXR is a large-scale chest X-ray dataset containing 473,057 radiographs with 206,563 associated reports. Tokens with frequency > 10 are retained, and reports are truncated/padded to 78 tokens to conform with CLIP's specifications. For a fair and consistent evaluation on the two datasets, we use the same data splits as employed in prior works [1, 3–8, 22, 23].

We evaluate report generation quality using standard natural language processing metrics: BLEU 1–4, METEOR, and ROUGE-L. All metrics are computed with a standard evaluation toolkit [24].

3.2   Implementation Details

The MHSAs and MHCAs in UniCrossAdapter use 64-dim features and 4 attention heads. For IU-Xray, the vision Transformer and report decoder have 3 layers each, while for MIMIC-CXR, we use 6 layers due to its larger size. To mitigate IU-Xray's limited data, we use a consolidated vocabulary combining both datasets, enabling more diverse word projections. We choose Adam as the optimizer and use a batch size of 16 for training. We employ an initial learning rate of 1e-5 and weight decays of 5e-5 and 4e-5 for IU-Xray and MIMIC-CXR, respectively. We also apply dropout for regularization with rates of 0.09 and 0.1 for the IU-Xray and MIMIC-CXR datasets, respectively.

3.3   Comparison with State-of-the-Art Methods

We compare against existing methods including R2Gen [1], SentSAT+KG [25], CMCL [26], $\mathcal{M}^2$ TR. PROGRESSIVE [27], CMN [3], PPKED [4], CMM+RL [5], XPRONET [6], DCL [8], M2KT [7], VLCI [28], RAMT [29], and PromptMRG [30]. As shown in Table 1, the proposed approach outperforms the best competing method by 2.4% in BLEU-2, 2.2% in BLEU-3, 1.4% in BLEU-4, 1.2% in BLEU-1, and 0.8% in METEOR on IU-Xray. While slightly lower in ROUGE-L compared to M2KT [7], our method remains the top performer overall. On the larger MIMIC-CXR dataset, our model also shows improvements of 0.8% in BLEU-3 and 0.7% in BLEU-4 compared to prior art, along with comparable BLEU-2 and ROUGE-L. As evidenced in previous work [1, 3–8, 25–30], gains on MIMIC-CXR are more marginal due to its scale. Overall, our approach achieves state-of-the-art or comparable performance on both IU-Xray and MIMIC-CXR datasets.

3.4   Ablation Study

We ablate key components of our model, UniCrossAdapter and CLIP encoders, to analyze their impact quantitatively (cf. Table 2). Removing either significantly

Table 2. Ablation results on the IU-Xray and MIMIC-CXR datasets. The best results are in **bold**. w/o denotes "without".

IU-Xray	BLEU-1	BLEU-2	BLEU-3	BLEU-4	ROUGE-L	METEOR
w/o UniCrossAdapter	0.302	0.201	0.146	0.109	0.375	0.154
w/o CLIP pre-training weights	0.450	0.298	0.208	0.147	0.357	0.188
Full model	**0.509**	**0.349**	**0.257**	**0.195**	**0.395**	**0.210**
MIMIC-CXR	BLEU-1	BLEU-2	BLEU-3	BLEU-4	ROUGE-L	METEOR
w/o UniCrossAdapter	0.087	0.055	0.038	0.028	0.226	0.077
w/o CLIP pre-training weights	0.351	0.196	0.118	0.077	0.250	0.118
Full model	**0.375**	**0.237**	**0.165**	**0.120**	**0.289**	**0.134**

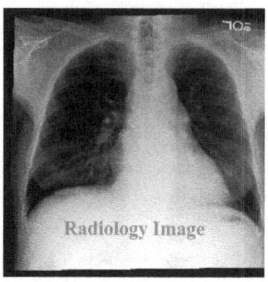

Radiology Image

Generation Results

Ours w/o UniCrossAdapter
the lungs are clear. there is no pleural effusion or pneumothorax. the lungs are clear. the lungs are. there is no pneumothorax. the lungs are. the right are.

Ours w/o CLIP pre-training weights
no change. the heart is normal. no the heart is normal. the lungs are clear. no pleural effusion or pneumothorax. the heart size is normal. the mediastinal and hilar contours are normal. no acute osseous abnormalities. no acute osseous abnormalities.

Ground Truth
the cardiomediastinal and hilar contours are normal. the lungs are well expanded and clear without focal consolidation pleural effusion or pneumothorax. mild degenerative changes are seen in the thoracic spine.

Ours
pa and lateral views of the chest were obtained. the lungs are clear. there is no focal consolidation pleural effusion or pneumothorax. the cardiomediastinal and hilar contours are unremarkable. there is no pulmonary edema. the cardiomediastinal silhouette is normal. there is no acute osseous abnormalities.

Fig. 2. Example of radiology report generation results on a test image from our model and its ablated variants. Ground truth words present in the generated reports are highlighted in color.

degrades performance, validating their efficacy. This suggests that CLIP's multimodal knowledge facilitates learning cross-modal semantic alignments.

Figure 2 shows example radiology reports generated by our full model and its ablated versions. In the absence of either UniCrossAdapter or CLIP pre-training weights, the model produces noticeably inferior results. Specifically, the generated results exhibit poor grammar and a high level of repetition. This demonstrates that introducing pre-trained cross-modal knowledge from CLIP into the task of radiology report generation proves highly effective for producing more comprehensive and fluent reports. Moreover, this also highlights the significance of vision-language alignment by our adapter method for the overall model. Furthermore, we observe that reports generated by our full model demonstrate a level of professionalism comparable to ground truths.

4   Conclusion

In this work, we propose leveraging CLIP for the task of automated radiology report generation. Recognizing the infeasibility of fully fine-tuning such a massive model, we introduce UniCrossAdapter, a parameter-efficient fine-tuning approach to adapt CLIP to this domain. Our experiments demonstrate state-of-the-art performance on two public benchmarks. Qualitative analysis shows our model is capable of generating coherent reports describing key clinical findings in medical images. This work illustrates the promise of large pre-trained multi-modal models for radiology report generation and introduces a method to make their adoption practical.

Acknowledgements.. This work is supported in part by the National Key Research and Development Program of China (2022ZD0160604), in part by the Natural Science Foundation of China (62101393/62176194), in part by the High-Performance Computing Platform of YZBSTCACC, and in part by MindSpore (https://www.mindspore.cn), a new deep learning framework.

Disclosure of Interests. The authors have no competing interests to declare that are relevant to the content of this paper.

References

1. Chen, Z., Song, Y., Chang, T.H., Wan, X.: Generating radiology reports via memory-driven transformer. In: 2020 Conference on Empirical Methods in Natural Language Processing, pp. 1439–1449 (2020)
2. Johnson, A.E., et al.: MIMIC-CXR-JPG, a large publicly available database of labeled chest radiographs. arXiv preprint arXiv:1901.07042 (2019)
3. Chen, Z., Shen, Y., Song, Y., Wan, X.: Cross-modal memory networks for radiology report generation. In: The Joint Conference of the 59th Annual Meeting of the Association for Computational Linguistics and the 11th International Joint Conference on Natural Language Processing, pp. 5904–5914 (2022)
4. Liu, F., Wu, X., Ge, S., Fan, W., Zou, Y.: Exploring and distilling posterior and prior knowledge for radiology report generation. In: IEEE/CVF Conference on Computer Vision and Pattern Recognition, pp. 13753–13762 (2021)
5. Qin, H., Song, Y.: Reinforced cross-modal alignment for radiology report generation. In: Findings of the Association for Computational Linguistics: ACL 2022, pp. 448–458 (2022)
6. Wang, J., Bhalerao, A., He, Y.: Cross-modal prototype driven network for radiology report generation. In: European Conference on Computer Vision, pp. 563–579 (2022)
7. Yang, S., Wu, X., Ge, S., Zheng, Z., Zhou, S.K., Xiao, L.: Radiology report generation with a learned knowledge base and multi-modal alignment. Med. Image Anal. **86**, 102798 (2023)
8. Li, M., Lin, B., Chen, Z., Lin, H., Liang, X., Chang, X.: Dynamic graph enhanced contrastive learning for chest X-ray report generation. In: IEEE/CVF Conference on Computer Vision and Pattern Recognition, pp. 3334–3343 (2023)

9. Demner-Fushman, D., et al.: Preparing a collection of radiology examinations for distribution and retrieval. J. Am. Med. Inform. Assoc. **23**(2), 304–310 (2016)

10. Sharma, P., Ding, N., Goodman, S., Soricut, R.: Conceptual captions: a cleaned, hypernymed, image alt-text dataset for automatic image captioning. In: The 56th Annual Meeting of the Association for Computational Linguistics, pp. 2556–2565 (2018)

11. Changpinyo, S., Sharma, P., Ding, N., Soricut, R.: Conceptual 12M: pushing web-scale image-text pre-training to recognize long-tail visual concepts. In: IEEE/CVF Conference on Computer Vision and Pattern Recognition, pp. 3558–3568 (2021)

12. Radford, A., et al.: Learning transferable visual models from natural language supervision. In: International Conference on Machine Learning, pp. 8748–8763 (2021)

13. Guo, D., Rush, A.M., Kim, Y.: Parameter-efficient transfer learning with diff pruning. In: The Joint Conference of the 59th Annual Meeting of the Association for Computational Linguistics and the 11th International Joint Conference on Natural Language Processing, pp. 4884–4896 (2021)

14. Gao, P., et al.: CLIP-Adapter: Better vision-language models with feature adapters. Int. J. Comput. Vision **132**(2), 581–595 (2024)

15. Chen, S., Ge, C., Tong, Z., Wang, J., Song, Y., Wang, J., Luo, P.: AdaptFormer: adapting vision Transformers for scalable visual recognition. In: Advances in Neural Information Processing Systems, pp. 16664–16678 (2022)

16. Zhou, K., Yang, J., Loy, C.C., Liu, Z.: Learning to prompt for vision-language models. Int. J. Comput. Vision **130**(9), 2337–2348 (2022)

17. Hu, E.J., Shen, Y., Wallis, P., Allen-Zhu, Z., Li, Y., Wang, S., Wang, L., Chen, W.: LoRA: Low-Rank adaptation of large language models. arXiv preprint arXiv:2106.09685 (2021)

18. Houlsby, N., et al.: Parameter-efficient transfer learning for NLP. In: International Conference on Machine Learning, pp. 2790–2799 (2019)

19. Rücklé, A., et al.: AdapterDrop: on the efficiency of adapters in transformers. In: 2021 Conference on Empirical Methods in Natural Language Processing, pp. 7930–7946 (2021)

20. Dosovitskiy, A., et al.: An image is worth 16×16 words: Transformers for image recognition at scale. arXiv preprint arXiv:2010.11929 (2020)

21. Vaswani, A., Shazeer, N., Parmar, N., Uszkoreit, J., Jones, L., Gomez, A.N., Kaiser, L., Polosukhin, I.: Attention is all you need. In: Advances in Neural Information Processing Systems, pp. 6000–6010 (2017)

22. Kong, M., Huang, Z., Kuang, K., Zhu, Q., Wu, F.: TranSQ: transformer-based semantic query for medical report generation. In: International Conference on Medical Image Computing and Computer Assisted Intervention, pp. 610–620 (2022)

23. Li, J., Li, S., Hu, Y., Tao, H.: A self-guided framework for radiology report generation. In: International Conference on Medical Image Computing and Computer Assisted Intervention, pp. 588–598 (2022)

24. Chen, X., Fang, H., Lin, T.Y., Vedantam, R., Gupta, S., Dollár, P., Zitnick, C.L.: Microsoft COCO captions: Data collection and evaluation server. arXiv preprint arXiv:1504.00325 (2015)

25. Zhang, Y., Wang, X., Xu, Z., Yu, Q., Yuille, A., Xu, D.: When radiology report generation meets knowledge graph. In: AAAI Conference on Artificial Intelligence, pp. 12910–12917 (2020)

26. Liu, F., Ge, S., Wu, X.: Competence-based multimodal curriculum learning for medical report generation. In: The Joint Conference of the 59th Annual Meeting of the Association for Computational Linguistics and the 11th International Joint Conference on Natural Language Processing, pp. 3001–3012 (2021)

27. Nooralahzadeh, F., Gonzalez, N.P., Frauenfelder, T., Fujimoto, K., Krauthammer, M.: Progressive transformer-based generation of radiology reports. In: Findings of the Association for Computational Linguistics: EMNLP 2021, pp. 2824–2832 (2021)

28. Chen, W., Liu, Y., Wang, C., Li, G., Zhu, J., Lin, L.: Visual-linguistic causal intervention for radiology report generation. arXiv preprint arXiv:2303.09117 (2023)

29. Zhang, K., et al.: Semi-supervised medical report generation via graph-guided hybrid feature consistency. IEEE Trans. Multimedia **26**, 904–915 (2024)

30. Jin, H., Che, H., Lin, Y., Chen, H.: PromptMRG: diagnosis-driven prompts for medical report generation. In: AAAI Conference on Artificial Intelligence, pp. 2607–2615 (2024)

TUMSyn: A Text-Guided Generalist Model for Customized Multimodal MR Image Synthesis

Yulin Wang[1,2], Honglin Xiong[2], Yi Xie[2], Jiameng Liu[2], Qian Wang[2,3], Qian Liu[1,4(✉)], and Dinggang Shen[2,3(✉)]

[1] School of Biomedical Engineering and State Key Laboratory of Digital Medical Engineering, Hainan University, Haikou 570228, China
qliu@hainanu.edu.cn

[2] School of Biomedical Engineering and State Key Laboratory of Advanced Medical Materials and Devices, ShanghaiTech University, Shanghai 201210, China
dgshen@shanghaitech.edu.cn

[3] Shanghai Clinical Research and Trial Center, Shanghai 201210, China

[4] Key Laboratory of Biomedical Engineering of Hainan Province, One Health Institute, Hainan University, Haikou 570228, China

Abstract. Multimodal magnetic resonance (MR) imaging has revolutionized our understanding of the human brain. However, various limitations in clinical scanning hinder the data acquisition process. Current medical image synthesis techniques, often designed for specific tasks or modalities, exhibit diminished performance when confronted with heterogeneous-source MRI data. Here we introduce a **T**ext-guided **U**niversal **M**R image **Syn**thesis (TUMSyn) generalist model to generate text-specified multimodal brain MRI sequences from any real-acquired sequences. By leveraging demographic data and imaging parameters as text prompts, TUMSyn achieves diverse cross-sequence synthesis tasks using a unified model. To enhance the efficacy of text features in steering synthesis, we pre-train a text encoder by using contrastive learning strategy to align and fuse image and text semantic information. Developed and evaluated on a multi-center dataset of over 20K brain MR image-text pairs with 7 structural MR contrasts, spanning almost entire age spectrum and various physical conditions, TUMSyn demonstrates comparable or exceeding performance compared to task-specific methods in both supervised and zero-shot settings, and the synthesized images exhibit accurate anatomical morphology suitable for various downstream clinical-related tasks. In summary, by incorporating text metadata into the image synthesis, the accuracy, versatility, and generalizability position TUMSyn as a powerful augmentative tool for conventional MRI systems, offering rapid and cost-effective acquisition of multi-sequence MR images for clinical and research applications.

Y. Wang and H. Xiong—These authors contributed equally to this work.

Z. Deng et al. (Eds.): MedAGI 2024, LNCS 15184, pp. 124–133, 2025.
https://doi.org/10.1007/978-3-031-73471-7_13

Keywords: Foundation Model · Multimodal MRI · MRI Synthesis · Super-resolution

1   Introduction

Magnetic resonance imaging (MRI) plays a pivotal role in neuroscience and clinical practice, enabling the exploration of the intricate structure and function of the human brain. However, clinical scanning constraints, including patient conditions and limited acquisition time, often lead to less sequences and suboptimal data quality. While several deep learning-based image synthesis methods [7,10,16–18] have been proposed to address these issues, their efficacy is often limited by task-specific and domain-specific training and inference paradigms. Consequently, these approaches exhibit suboptimal performance when applied to real-world clinical and research scenarios characterized by heterogeneous data.

Recent advances have sparked an increase in interest in exploring foundation models [11,12,20]. Built on large-scale and diverse datasets, these models exhibit flexibility in tackling multiple modalities, generating expressive outputs, and swiftly adapting to downstream tasks not explicitly defined by the training datasets by leveraging their mastered knowledge.

Drawing inspiration from the multimodal processing capabilities and robust generalizability of foundation models, here we present a **T**ext-guided **U**niversal **MR** image **Syn**thesis (TUMSyn) generalist model to synthesize neuroimages across nearly entire lifespan and multiple structural MRI modalities from any available MRI scan. By incorporating textual imaging parameters and demographic information into the image generation process, TUMSyn enables a single model to perform all cross-sequence synthesis tasks guided by text prompts, and achieves zero-shot generalization to novel data domains and tasks.

To achieve these capabilities, we embark on constructing a comprehensive brain MRI dataset, including over 20K 3D scans from public repositories and proprietary sources across multiple institutions. This extensive dataset incorporates seven common structural MRI modalities and represents a diverse population aged 2 to 100+, including healthy individuals and patients with various conditions. Furthermore, we pre-train a text encoder by leveraging the contrastive learning strategy [14,19] to extract image-aligned text features to enhance the efficacy of text prompts in guiding image synthesis.

To assess the accuracy, versatility, and generalizability of our model, we conduct experiments on multi-center MR scans of controls and patients with brain tumors and Alzheimer's disease. The remarkable generated results and their effectiveness in downstream tasks demonstrate our model's proficiency in customized sequence synthesis and seamless adapting to various MR data domains prevalent across hospitals and institutions, with the promise of streamlining clinical workflows and reducing healthcare costs by augmenting acquired MR scan(s).

2   Method

The TUMSyn framework, illustrated in Fig. 1, achieves general cross-sequence image synthesis through a two-stage process. In the first stage, we pre-train a text encoder to learn the relationships between MR image representations and their corresponding demographic and imaging information. The second stage employs the frozen text encoder as the prompt producer to flexibly guide target images generation using any of the available sequences.

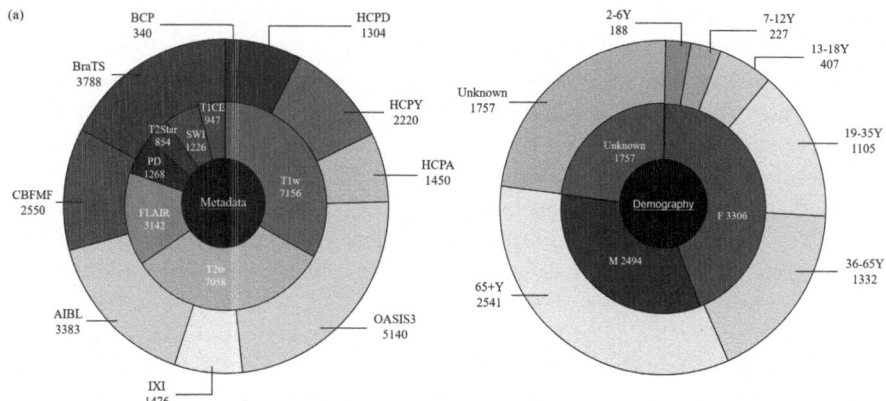

(b) **Age:**{34.1}; **Gender:**{M}; **Scanner:**{3.0 Siemens}; **Modality:**{T1w}; **Voxel size:**{(0.8, 0.8, 0.8)}; **Imaging parameter TR(ms), TE(ms), TI(ms), and FA(degree):**{(2500.0, 2.2, 1000.0, 8.0)}

Fig. 1. Overview of the training set of brain MRI dataset. (a) The left plot illustrates the sample numbers of each dataset and MR modality, and the right one presents the sample numbers of each age group and gender. (b) A sample of the text prompt.

2.1   Data Collection

To ensure the versatility and generalizability of our model, our brain MRI multi-modal dataset comprises over 20K 3D brain MRI scans from diverse global institutions, including OASIS [8], HCP [15], IXI [1], BCP [6], ADNI [13], AIBL [5], BraTS2021 [2], and a in-built dataset (CBMFM). And the HCP database contains three age groups, *i.e.*, HCP Young (HCPY), HCP Development (HCPD), and HCP Aging (HCPA). This comprehensive collection encompasses seven prevalent structural MRI modalities, *i.e.*, T1-weighted (T1w), T2-weighted (T2w), Fluid Attenuated Inversion Recovery (FLAIR), Susceptibility Weighted Imaging (SWI), T2 Star, Proton Density (PD), and Contrast-Enhanced T1w (T1CE), representing an entire lifespan cohort that includes healthy individuals and patients with neurodegenerative and neurodevelopment disorders and brain tumors. The detailed description of the training set is illustrated in Fig. 1(a), and we reserved the ADNI dataset for external validation experiments to ensure a robust evaluation of our model's generalizability. To ensure the model can

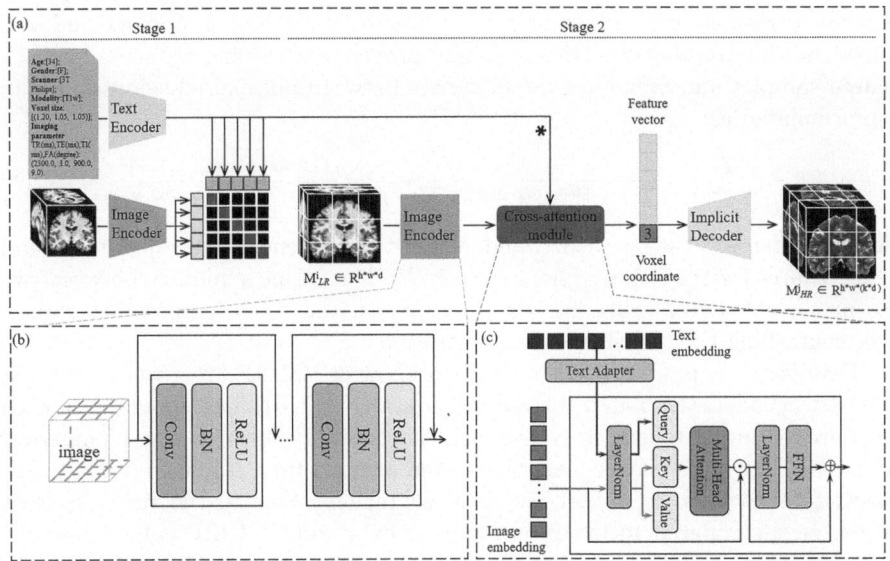

Fig. 2. Overview of TUMSyn. (a) The pipeline of prompt-guided MR image synthesis and super-resolution, which includes pre-training image-text alignment (Stage 1) and universal MR image SR and cross-sequence synthesis (Stage 2). i and j represent different modalities. (b) Model architecture of image encoder. (c) The model architecture of image-text cross-attention module.

adapt to real clinically heterogeneous data as well as learn general MR image features, we simplify the image pre-processing steps as follows: (i) registering multiple modalities of each subject, and (ii) skull-stripping.

Concurrently, the corresponding textual metadata for each scan is also an important part of our dataset. A sample of our template is shown in Fig. 1(b), where TR, TE, TI, and FA represent repetition time, echo time, inversion time, and flip angle, respectively. These parameters significantly influence the contrast and signal intensity of the scanned images. All imaging parameters and demographic information are derived from DICOM header files, official websites, and demographic statements. In cases of missing information, we consistently use 'None' placeholders. Note that information about the target modality is invariably present in all text prompts.

2.2 Pre-trained Text Encoder for Enhanced Feature Extraction

To enable precise guidance of downstream image synthesis tasks by textual prompts, we introduce a contrastive language-image pre-training (CLIP) model dedicated to brain MR images (BMLIP) in the first stage. Our goal is to equip the text encoder with the capability to extract metadata embeddings corresponding to images. The overall framework is depicted in Stage 1 of Fig. 2(a). Specifically, given a batch of text-image pairs as inputs, BMLIP learns multimodal feature correspondences in latent space through joint training of image and text encoders. The training objective is to minimize the embedding distances between paired samples and maximize the distances between non-paired samples. It can be formulated as:

$$\mathcal{L} = -\frac{1}{2N} \sum_{i=1}^{N} \left[\log \frac{\exp(\cos(\mathbf{v}_i, \mathbf{t}_i)/\tau)}{\sum j = 1^N \exp(\cos(\mathbf{v}_i, \mathbf{t}_j)/\tau)} + \log \frac{\exp(\cos(\mathbf{t}_i, \mathbf{v}_i)/\tau)}{\sum j = 1^N \exp(\cos(\mathbf{t}_i, \mathbf{v}_j)/\tau)} \right] \quad (1)$$

Where N denotes the batch size, and $\mathbf{v}_i$ and $\mathbf{t}_i$ represent the encoded image and text features for the i-th sample, respectively. The cosine similarity between two features is calculated using the $\cos(\cdot, \cdot)$ operator, and τ is a temperature parameter controlling the distribution concentration.

Details of the model structure and workflow of BMLIP are given below. For the text encoder, templated textual metadata (Fig. 1(b)) corresponding to each scan are fed into a tokenizer, where each character is converted into a numerical identifier. The numerical sequences are then passed into a standard transformer-based text encoder to obtain text representations. Due to the richer content in our text compared to the prompts used by standard CLIP [14], we set the encoded length to 90 tokens. For the image encoder, recognizing MRI scans are typically 3D and the critical importance of inter-slice structural information for diagnostic and downstream image analysis tasks, we modify the ViT-B/16 architecture used in standard CLIP designed for 2D natural images to accommodate 3D MR images. To manage the computational demands of 3D image processing, we first downscale the entire input image to half of its original size and then crop all images to $96 \times 96 \times 96$ resolutions. This approach preserves semantic information in the images, ensuring effective semantic matching between text and images during training. Upon completion of BMLIP training, we only apply the text encoder to the subsequent stage.

2.3   Tailored MRI Synthesis and Super-Resolution Guided by Text Prompt

The cross-sequence image synthesis model, illustrated in Stage 2 of Fig. 2(a), aims to generate tailored target MRI scans from available images of varying MR modalities, resolutions, and orientations, guided by text prompts. This process begins with encoding input pairs of text and image patches ($60 \times 60 \times 60$) using a convolutional neural network (CNN)-based image encoder and a frozen pre-trained text encoder to extract multi-modality features, respectively. To accommodate super-resolution requirements, we employ a modified ResNet-34 without downsampling as our image encoder (Fig. 2(b)). Textual MR knowledge is incorporated into the synthesis process through a cross-attention module (Fig. 2(c)). The cross-modality integration process begins with a text adapter that updates text embeddings to align with our specific task requirements. Subsequently, multi-head attention mechanisms fuse these adapted text embeddings with image features, generating multi-modality representations enriched by textual information. To ensure the versatility of our model, we implement an image decoder capable of arbitrary resolution upsampling. This is achieved through the Local Implicit Image Function (LIIF) [4], which adds high-resolution image coordinates to the feature vector around the low-resolution coordinates as input and predicts the intensity value at a given continuous coordinate using an implicit decoder. Finally, we obtain an image with tailored demographic and imaging parameter information specified in the text prompt. In this stage, we employ pixel loss as supervision to synthesize images.

3 Experiment

3.1 Implementation Details

TUMSyn is implemented using PyTorch 1.12.1 and trained on a server equipped with a Nvidia A100 GPU. We adopt the Adam optimizer for both model training stages. The learning rate (LR) in the first stage increases from 0 to 0.0005 for model warm-up, and then gradually decreases to 0, while the LR in the second stage is set to 0.0001 initially and decays by half every 100 epochs. The training epochs for models in both stages are 100 and 300, respectively.

3.2 Comparative Evaluation of Accuracy and Versatility in Image Synthesis

In this experiment, we aim to demonstrate the model's capability to accurately and uniformly generate target images under the guidance of text prompts. First, we compare TUMSyn with several state-of-the-art methods and perform in-depth quantitative assessments across seven typical tasks, as outlined in Table 1. SC-GAN [9] is a task-specific image mapping method. The one-hot model and BiomedCLIP model share the identical synthesis model architecture, but one-hot model uses numerical labels to replace the text prompts and BiomedCLIP model uses Biomedclip [19] to produce text embeddings. The PSNR and SSIM scores reveal that TUMSyn consistently exhibits superior performance across all the tasks. Compared to SC-GAN, our findings highlight the effectiveness of generic feature representations learning from abundant data. When compared to BiomedCLIP, the results underscore the necessity of specifically pre-trained foundational models in the medical domain. The comparison results with the one-hot method emphasize the accuracy of leveraging text prompts to steer the image translation process. We also perform a comparative analysis of the qualitative outcomes between TUMSyn and the aforementioned competitors on multiple tasks (Fig. 3). SynthSR [7] is further enrolled, which is a general MRI image synthesis algorithm that specializes in translating diverse MRI sequences into T1w sequences with isotropic voxel sizes.

Notably, TUMSyn adeptly achieves the best results in all cases. Rows 1 to 4 illustrate the capacity of TUMSyn to generate modalities tailored to the specified resolution, MR contrast, and voxel intensity as prompted by the text, irrespective of the imaging parameters of the input sequences. Subsequently, the final two rows demonstrate the remarkable ability of BMLIP to capture and retain

Table 1. Universality and accuracy comparison with different methods on the test set

		AIBL PD->FLAIR	HCPD T1->T2	BRATS FLAIR->T1CE	CBMFM T2->FLAIR	IXI T1->PD	OASIS T1->SWI	OASIS T2->T2S
SC-GAN[9]	PSNR	25.05 ± 0.98	28.24 ± 0.43	24.11 ± 1.52	28.27 ± 1.82	30.33 ± 0.87	25.48 ± 1.02	28.48 ± 0.99
	SSIM	0.903 ± 0.012	0.940 ± 0.005	0.871 ± 0.015	0.958 ± 0.020	0.961 ± 0.003	0.851 ± 0.019	0.939 ± 0.011
BiomedCLIP[19]	PSNR	27.26 ± 1.22	28.76 ± 0.61	26.37 ± 1.48	30.91 ± 2.26	31.29 ± 1.10	25.63 ± 1.15	28.48 ± 1.14
	SSIM	0.937 ± 0.012	0.943 ± 0.007	0.913 ± 0.013	0.975 ± 0.019	0.967 ± 0.007	0.855 ± 0.028	0.938 ± 0.016
One-hot	PSNR	25.58 ± 0.97	27.27 ± 0.67	24.57 ± 1.23	28.63 ± 1.82	28.57 ± 0.51	21.87 ± 0.92	25.12 ± 0.88
	SSIM	0.919 ± 0.010	0.930 ± 0.008	0.905 ± 0.014	0.966 ± 0.019	0.635 ± 0.017	0.849 ± 0.030	0.911 ± 0.015
TUMSyn	PSNR	**27.43 ± 1.17**	**29.02 ± 0.68**	**26.78 ± 1.67**	**31.13 ± 2.43**	**31.78 ± 0.88**	**25.63 ± 1.02**	**28.49 ± 1.17**
	SSIM	**0.938 ± 0.010**	**0.945 ± 0.007**	**0.915 ± 0.013**	**0.976 ± 0.019**	**0.969 ± 0.007**	**0.856 ± 0.028**	**0.939 ± 0.016**

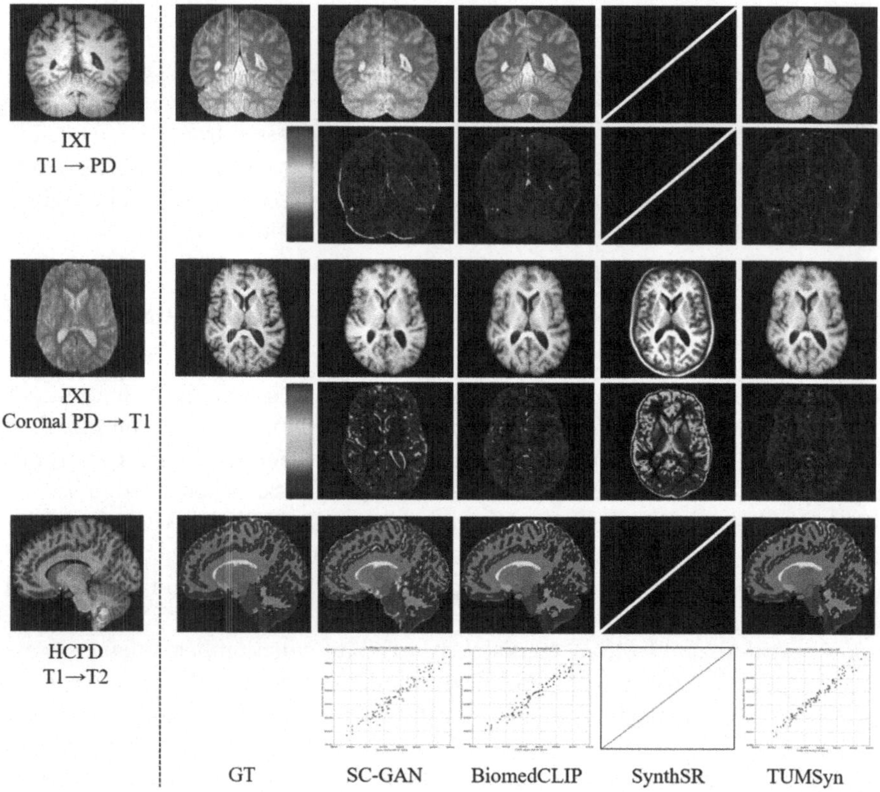

Fig. 3. Visualization of synthesized images from heterogeneous input data. Row 1 and 2 show synthetic PD images from real-acquired T1w and corresponding error maps compared to the ground truth in the IXI dataset, respectively. Row 3 and 4 present T1w synthetic results from simulated coronal PD sequence with 3.2mm slice spacing and corresponding error maps compared to ground truth, respectively, and we present these results in the orthogonal view to illustrate the performance of joint image super-resolution and cross-sequence synthesis. Row 5 and 6 present whole-brain segmentation results obtained using SynthSeg+ and scatter plots of cortex volumes derived from real-acquired T1w and synthetic T2w images, respectively.

Table 2. Hippocampal volumetry of AD, MCI, and NC. The rows indicate the average (Avg.) volume of ROI for different cognitive stages.

	Real T1w	SC-GAN	BiomedCLIP	SynthSR	TUMSyn
Avg. AD (mm^3)	3280.11	3326.58	3560.73	3419.84	3391.97
Avg. MCI (mm^3)	3614.33	3664.55	3957.30	3817.15	3770.19
Avg. NC (mm^3)	4030.67	4037.89	4237.52	4061.15	4040.37

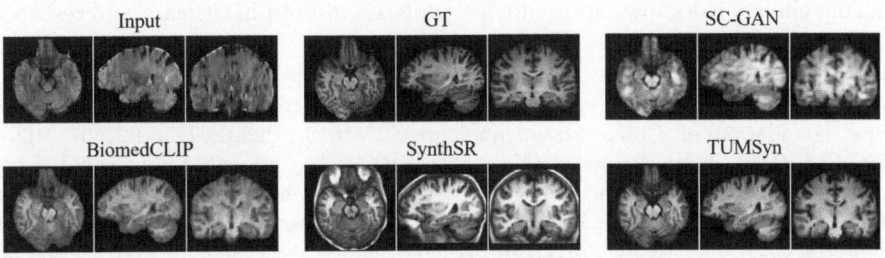

Fig. 4. Joint image synthesis and SR results of a MCI subject from ADNI datasets. The hippocampal segmentation results obtained by Synthseg+ are overlaid on the generated T1w images from 5mm axial FLAIR.

brain anatomical structures and generate brain tissues, and the segmentation results obtained from Synthseg+ [3] reaffirm the compatibility of its generated outputs with existing medical image analysis tools.

3.3   Zero-Shot Detecting AD-Induced Brain Region Atrophy

The full potential of a foundation model cannot be realized without sufficient generalization. In this experiment, we evaluate TUMSyn's zero-shot performance using an out-of-distribution dataset ADNI, assessing its ability to generate lesion images that reflect the effects of cognitive impairment on hippocampal shape and volume. The hippocampus is a key biomarker of neurodegenerative diseases, and typically exhibits progressive volume reduction as the disease advances. Table 2 presents the average hippocampal volumes for Alzheimer's Disease (AD), Mild Cognitive Impairment (MCI), and Normal Control (NC), as estimated by each method. Volumes derived from real T1w images serve as the ground truth for each disease stage.

In comparison to other methods, TUMSyn demonstrates robust lesion volume delineation capabilities and strong discriminative power across the three disease stages, approaching the performance of real T1w images and SC-GAN trained on this dataset. Figure 4 illustrates the T1w sequence outputs from various methods and the resulting hippocampal segmentation for a representative MCI patient scan. Qualitatively, TUMSyn exhibits the superior recovery of high-frequency details, facilitating accurate hippocampal segmentation when used with Synthseg+.

4 Conclusion

In this study, we present TUMSyn, a text-guided universal model for the unified synthesis of tailored structural MRI modalities, using any available MRI sequences. Developed and evaluated on a diverse-source dataset comprising over 20K 3D MRI scans, TUMSyn demonstrates promising synthesis accuracy in producing diagnostically relevant sequences and remarkable adaptability to novel data domains. Altogether, TUMSyn holds significant potential for optimizing clinical workflows and reducing expenses in both healthcare and research settings.

Acknowledgement. This work was supported in part by National Natural Science Foundation of China (grant numbers 62131015, U23A20295, 62250710165), the STI 2030-Major Projects (No. 2022ZD0209000), Shanghai Municipal Central Guided Local Science and Technology Development Fund (grant number YDZX20233100001001), The Key R&D Program of Guangdong Province, China (grant numbers 2023B0303040001, 2021B0101420006), and Science and Technology special fund of Hainan Province (grant number KJRC2023B06).

Disclosure of Interests. The authors declare no competing interests.

References

1. IXI dataset. https://brain-development.org/ixi-dataset/
2. Baid, U., et al.: The rsna-asnr-miccai brats 2021 benchmark on brain tumor segmentation and radiogenomic classification. arXiv preprint arXiv:2107.02314 (2021)
3. Billot, B., et al.: SynthSeg: Segmentation of brain MRI scans of any contrast and resolution without retraining. Med. Image Anal. **86**, 102789 (2023)
4. Chen, Y., Liu, S., Wang, X.: Learning continuous image representation with local implicit image function. In: Proceedings of the IEEE/CVF Conference on Computer Vision and Pattern Recognition, pp. 8628–8638 (2021)
5. Ellis, K.A., et al.: The australian imaging, biomarkers and lifestyle (AIBL) study of aging: methodology and baseline characteristics of 1112 individuals recruited for a longitudinal study of alzheimer's disease. Int. Psychogeriatr. **21**(4), 672–687 (2009)
6. Howell, B.R., et al.: The UNC/UMN baby connectome project (BCP): an overview of the study design and protocol development. Neuroimage **185**, 891–905 (2019)
7. Iglesias, J.E., et al.: SynthSR: a public AI tool to turn heterogeneous clinical brain scans into high-resolution t1-weighted images for 3D morphometry. Sci. Adv. **9**(5), eadd3607 (2023)
8. LaMontagne, P.J., et al.: Oasis-3: longitudinal neuroimaging, clinical, and cognitive dataset for normal aging and alzheimer disease. MedRxiv, pp. 2019–2012 (2019)
9. Lan, H., Initiative, A.D.N., Toga, A.W., Sepehrband, F.: Three-dimensional self-attention conditional GAN with spectral normalization for multimodal neuroimaging synthesis. Magn. Reson. Med. **86**(3), 1718–1733 (2021)
10. Liu, J., Pasumarthi, S., Duffy, B., Gong, E., Datta, K., Zaharchuk, G.: One model to synthesize them all: multi-contrast multi-scale transformer for missing data imputation. IEEE Trans. Med. Imaging (2023)

11. Ma, J., Wang, B.: Towards foundation models of biological image segmentation. Nat. Methods **20**(7), 953–955 (2023)
12. Moor, M., et al.: Foundation models for generalist medical artificial intelligence. Nature **616**(7956), 259–265 (2023)
13. Petersen, R.C., et al.: Alzheimer's disease neuroimaging initiative (ADNI): clinical characterization. Neurology **74**(3), 201–209 (2010)
14. Radford, A., et al.: Learning transferable visual models from natural language supervision. In: International Conference on Machine Learning, pp. 8748–8763. PMLR (2021)
15. Van Essen, D.C., et al.: The WU-Minn human connectome project: an overview. Neuroimage **80**, 62–79 (2013)
16. Wang, Y., et al.: A unified hybrid transformer for joint MRI sequences super-resolution and missing data imputation. Phys. Med. Biol. (2023)
17. Wang, Y., et al.: Deep learning-based 3D MRI contrast-enhanced synthesis from a 2D noncontrast t2flair sequence. Med. Phys. **49**(7), 4478–4493 (2022)
18. Wu, Q., et al.: An arbitrary scale super-resolution approach for 3D MR images via implicit neural representation. IEEE J. Biomed. Health Inform. **27**(2), 1004–1015 (2022)
19. Zhang, S., et al.: Large-scale domain-specific pretraining for biomedical vision-language processing. arXiv preprint arXiv:2303.00915 **2**(3), 6 (2023)
20. Zhou, Y., et al.: A foundation model for generalizable disease detection from retinal images. Nature **622**(7981), 156–163 (2023)

SAMU: An Efficient and Promptable Foundation Model for Medical Image Segmentation

Joseph Bae[✉], Xueqi Guo, Halid Yerebakan, Yoshihisa Shinagawa, and Sepehr Farhand

Siemens Healthineers, Malvern, PA19355, USA
{joseph.bae,sepehr.farhand}@siemens-healthineers.com

Abstract. Segmentation of 3D medical images is a labor-intensive task with important clinical applications. Recently, foundation models for image segmentation have received significant interest. Specifically, many works have proposed methods for the adaptation of promptable natural image foundation models to medical image segmentation. However, the shift to 3D volumes from 2D natural images has proven difficult, and many approaches have limited real-world clinical applicability due to large model sizes and corresponding heavy computational requirements. Here, we present an original model for generalized, promptable 3D medical image segmentation. Our approach leverages a lightweight convolutional backbone while simultaneously integrating information from single-point prompts at multiple spatial resolutions. Our approach dramatically reduces the computational burden for promptable segmentation while also outperforming similar recent works on a diverse dataset of 98,699 image-mask pairs from CT and MRI datasets.

Keywords: Foundation Models · Segmentation · Prompting

1 Introduction

Medical image segmentation is an important task for the diagnosis, treatment, and management of human diseases. Automated delineation of abnormalities can prevent missed diagnoses and enable early detection of pathology. Interventional treatments including radiotherapy for malignancies rely heavily on accurate healthy organ segmentation in order to minimize damage to non-targeted structures. Finally, longitudinal contouring of pathologies can inform treatment decisions based on changes in disease presentation. However, manual segmentation of 3D medical images is a labor-intensive process in routine clinical workflows. While recent foundation models have been presented for generalized segmentation of both natural and medical images, the massive models proposed are difficult to train, finetune, or use in low-resource settings including for individual hospitals or clinics. This is particularly salient when many foundation models trained on natural images must be finetuned or completely

Z. Deng et al. (Eds.): MedAGI 2024, LNCS 15184, pp. 134–142, 2025.
https://doi.org/10.1007/978-3-031-73471-7_14

re-trained from scratch on medical imaging datasets in order to overcome the large domain shift from natural to medical imaging. Inspired by the Segment Anything Model (SAM) [9] created by Meta AI for natural image segmentation, we propose a lightweight, CNN-based architecture for promptable segmentation of 3D medical images. Our approach is adaptable to multiple anatomical sites and imaging modalities while requiring significantly fewer computational resources and outperforming recently proposed Vision Transformer [5] (ViT)-based methods.

SAM is among the most widely used foundation models for segmentation of natural images and can leverage a single-point prompt as input. However, it has been observed that the original implementation of SAM has difficulty generalizing to medical imaging tasks [11]. As a result, multiple strategies have been proposed to adapt SAM to medical images with varying success [4,10,13,14]. Currently, most of these works have focused primarily on 2D image segmentation [4,10], with 3D volumes being segmented on a slice-by-slice basis. This approach is generally ineffective at producing high-fidelity segmentations with spatial consistency across slices, and also requires modeling every individual image slice with associated high-computation costs and a need for per-slice prompts. SAM-Med3D [14] attempted to address these problems by training a SAM model using 3D ViT blocks on a large dataset of 131,000 mask-image pairs, thereby natively accommodating 3D inputs with a 3D architecture. However, the 100 million trainable parameters in the model make it difficult to fine-tune and result in relatively slow inference times. FastSAM3D [13] was proposed to reduce this computational burden through a knowledge distillation process in which a lightweight TinyViT architecture was trained as a student model with SAM-Med3D as a teacher. Additionally, FastSAM3D incorporated 3D sparse flash attention to further improve model efficiency. While FastSAM3D was able to effectively reduce the number of trainable parameters required for promptable segmentation to 53 million, the model is still large enough to be potentially prohibitive for hospitals and clinics with low computational resources. Further, both SAM-Med3D and FastSAM3D perform relatively poorly in the one-point prompt segmentation setting.

We focus here primarily on the single-point prompt segmentation task for the following reasons. First, in busy clinical settings, point annotation of only one slice in a 3D volume is more efficient and practical than requiring multiple annotations. Second, many previous implementations of multiple-point prompting are ambiguous, and may not be a realistic reflection of presumed real-world usage. For instance, in some implementations, each point beyond the first is chosen only within the false-negative region, implying a gradual and interactive refinement of model outputs in inference. This can prohibitively increase inference times due to repetitive modeling of a single image and requires significantly more human resources compared to single-point prompting.

In this work, we propose a lightweight model inspired by **SAM** and UNet (SAMU) for promptable 3D medical image segmentation. Our approach is motivated by a need for a computationally inexpensive model for one-point volu-

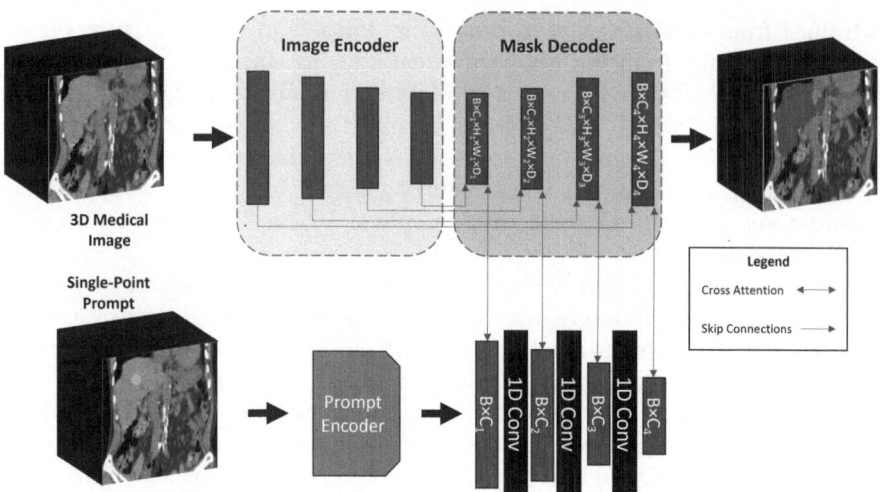

Fig. 1. SAMU Architecture. Shown is the proposed architecture for our promptable segmentation framework. SAMU leverages a UNet backbone for image encoding and mask decoding. The prompt encoding module from SAM is utilized to generate feature representations of single-point prompts. Cross-attention is performed at multiple spatial resolutions for prompt supervision of mask decoding.

metric segmentation and is trained and evaluated on a large dataset of 98,699 image-mask pairs (N = 2,703 subjects) spanning CT and MRI modalities. Our approach is comprised of the following contributions:

1. SAMU enables 3D medical image segmentation with significantly fewer parameters than other 3D approaches.
2. SAMU incorporates multi-scale prompt encoding to more thoroughly exploit positional information from single-point prompts.
3. SAMU outperforms previously published baselines on a diverse dataset of >90,000 mask-image pairs.

2   Methodology

We first describe the promptable segmentation framework popularized by SAM. We then elaborate upon our proposed architecture and highlight key innovations. An overview of our approach is presented in Fig. 1.

2.1   SAM Overview

SAM is composed of an image encoder, a prompt encoder, and a mask decoder. The image encoder is a ViT-H model pre-trained via a masked auto-encoder strategy [7,16] that leverages sequential attention and multi-layer perceptron blocks to extract image representations for the segmentation task. The original SAM prompt encoder flexibly allows point, bounding box, mask, and text prompts. For point prompts, a learned representation is created by concatenating a positional encoding vector with learnable tokens for segmentation. The mask decoder performs cross-attention between prompt and image embeddings before upsampling image embeddings to an output mask prediction.

2.2 Proposed SAMU

SAMU Encoder-Decoder Backbone. Unlike SAM, SAMU integrates both image encoder and mask decoder into a single UNet-like architecture. Specifically, for the encoding branch, SAMU utilizes 3D convolutions combined with max-pooling operations for feature downscaling. In the decoder branch, 3D transposed convolutions are used to recover the spatial dimensions of the input. Further, skip connections between corresponding spatial dimensions in the encoder and decoder branches of SAMU are implemented. This architecture allows the network to leverage the multi-scale feature information present at different spatial resolutions, a capability demonstrated to be significant in other medical imaging applications of UNet. Further, this framework is significantly less computationally expensive when compared to ViT encoders.

SAMU One-Point Prompt Encoding. For each volumetric 3D medical image studied, a single-point prompt is encoded into a positional embedding with concatenated learnable tokens as described in SAM. We discard the bounding box, mask, and text prompt encoding modules of SAM. Each learned prompt representation is incorporated into the decoder branch of SAMU at multiple resolutions. Specifically, at each spatial resolution in the UNet decoder, cross-attention is performed between the prompt feature vector and the image embedding. To align the dimensions of the prompt feature vector with the channel dimension of each image embedding, 1D convolutions are performed. Therefore, the prompt feature vector p_n is of $B \times C_n$ dimensionality where B is the batch size and C_n is the number of channels of the image representation r_n at the nth level of the UNet backbone. As a result of these steps, important localizing information present in the single-point prompts is provided to the network at multiple spatial resolutions, ensuring that the model can adequately learn which regions to segment during inference.

3 Experiments and Results

3.1 Datasets and Implementation

We studied one-point segmentation of a variety of anatomical and pathological targets of interest. For each volumetric medical image, a single-point prompt was randomly chosen within the entire 3D ground truth of each target volumetric mask. 15 abdominal organs from the AMOS [8] abdominal CT/MRI dataset

(N = 360), 3 tumor regions from the Brain Tumor Segmentation 2021 (BraTS) [1,2,12] dataset (N = 1,251), and 117 anatomical structures from the Total Segmentator (TotalSeg) [15] dataset (N = 1,092) were studied. Data partitioning was performed in the same manner as in SAM-Med3D [14] and FastSAM3D [13] for fair comparisons. The official training split for AMOS was further randomly divided for training and validation while the official validation split was used for testing. The training split of BraTS was randomly divided into training, validation, and testing splits at a 70/10/20 ratio. Individual MRI sequences in BraTS were treated as independent samples, but all images for a given patient could only appear in one of the train, validation, or test splits in order to avoid data leakage. Training, validation, and testing splits were used as described by the Total Segmentator dataset. A single instance of SAMU was trained concurrently on all three training datasets and evaluated on all three test datasets; individual models were not independently trained for each dataset and modality. Prior to image input, each volume was z-score normalized and cropped to a $128 \times 128 \times 128$ voxel patch as described in SAM-Med3D [14] and FastSAM3D [13]. Random flipping was used for data augmentation. The Dice score metric was used to compare segmentation quality.

All models were trained on the same single NVIDIA RTX 6000 Ada Generation GPU. SAMU was trained with a batch size of 16 and optimized with AdamW with a learning rate of 0.001. Pre-trained weights for Med-SAM3D and FastSAM3D were obtained from their respective authors.

Table 1. Dice Scores of Segmentation Results

		AMOS	BraTS	TotalSeg
	SAM [9]	0.049	0.108	0.202
2D	SAM-Med2D [4]	0.097	0.013	0.008
	MedSAM [10]	0.004	0.008	0.006
3D	SAM-Med3D [14]	0.453	0.365	0.334
	FastSAM3D [13]	0.307	0.315	0.242
	SAMU (ours)	**0.677**	**0.434**	**0.756**

3.2   Segmentation Results

Table 1 displays Dice scores results for SAMU as well as 2D and 3D baselines across all datasets studied when using a single-point prompt. Note that 2D results are directly reported in [13]; we were unable to replicate each 2D approach due to limited resources. Furthermore, each 2D method required slice-by-slice rather than full volume inference, therefore also requiring a single-point prompt per slice rather than a single-point prompt per image volume. Despite this advantage, SAMU and other 3D approaches still outperform their 2D counterparts on medical imaging datasets. Furthermore, SAMU outperforms SAM-Med3D and FastSAM3D in the single-point prompt evaluation schema across all datasets.

Qualitative results from SAM, SAM-Med3D, FastSAM3D, and SAMU are presented in Fig. 2. The original SAM model was frequently unable to achieve satisfactory performance on medical imaging tasks. Additionally, SAM-Med3D and FastSAM3D seem to heavily rely on prompt location and do not always produce quality segmentations conforming to anatomical structures. In contrast, SAMU produces segmentations that more accurately delineate targets of interest. Note that the same single-point prompt was provided to each model studied.

Table 2. Dice Scores for Ablation Studies

	AMOS	BraTS	TotalSeg
Baseline 1	0.586	0.358	0.694
Baseline 2	0.643	0.378	0.733
Baseline 3	0.628	0.400	0.676
SAMU (ours)	**0.677**	**0.434**	**0.756**

3.3 Ablation Experiments

To evaluate the impact of different components of SAMU, we compared model performance to that of a variant with no prompt provided (Baseline 1), a version of SAMU in which prompt cross-attention is only performed at the bottom layer of the UNet architecture (Baseline 2), and a SAMU variant in which cross-attention is applied to both image embeddings and skip-connected residual feature maps (Baseline 3). The results are shown in Table 2. As might be expected, the addition of prompts (Baselines 2 and 3, SAMU) improves model performance. Further, the absence of multi-scale prompt information seems to heavily reduce performance (Baseline 2 vs. SAMU), as does cross attention between prompts and residual features (Baseline 3 vs. SAMU).

3.4 Computational Requirements

Table 3 displays the resource requirements of SAMU when compared with SAM, SAM-Med3D, and FastSAM3D. In addition to using substantially fewer parameters, SAMU is faster in inference. It should be noted that 3D models are significantly faster for 3D medical imaging inference as 2D SAM approaches require per-slice inference along with input of per-slice single-point prompts rather than per-volume prompts.

Table 3. Computational Requirements

	Params	Inference Time (s)
SAM [9]	636 M	52.521
SAM-Med3D [14]	100 M	1.188
FASTSAM3D [13]	53 M	0.280
SAMU (ours)	**12 M**	**0.049**

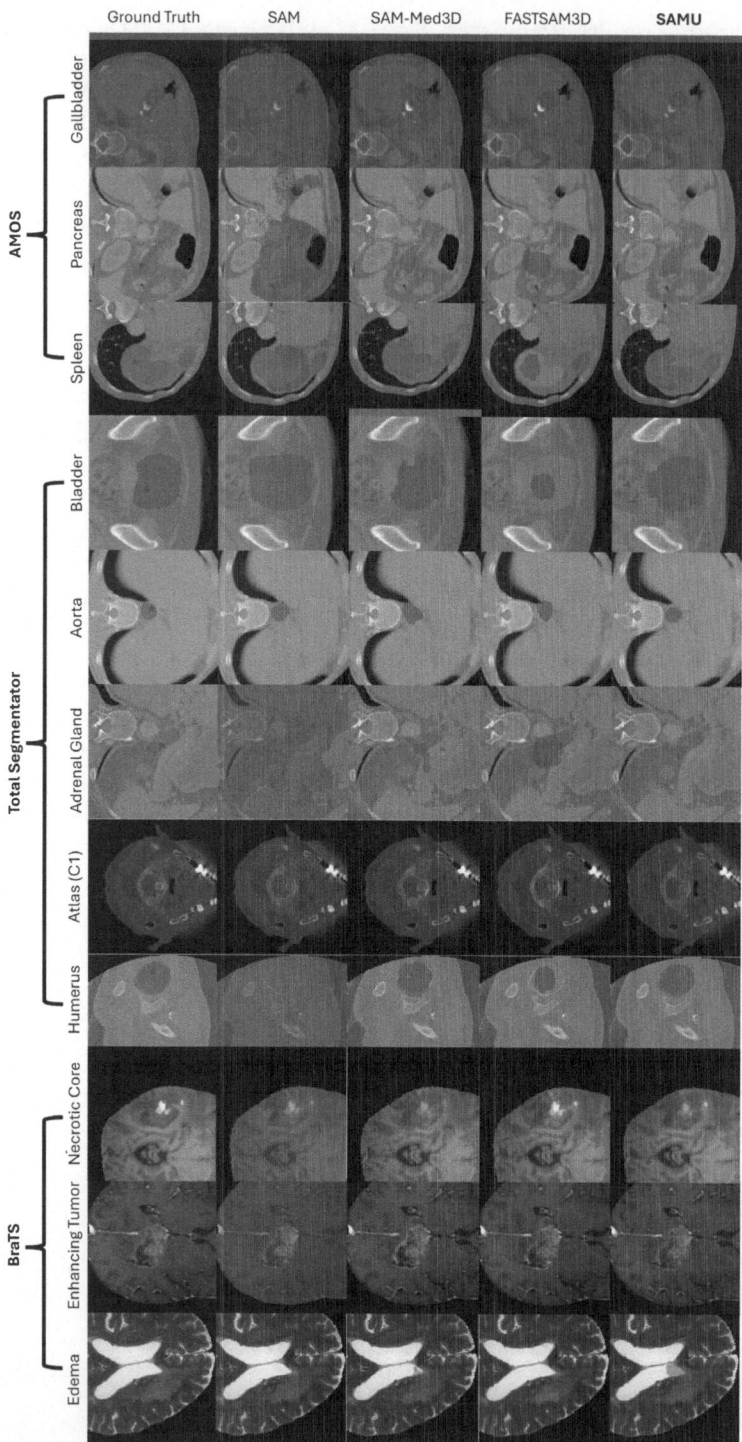

Fig. 2. Qualitative Results. Shown are example segmentations for the studied approaches. Single-point prompts were identically chosen for each method (green star).

4   Discussion and Conclusion

Here we present SAMU, a lightweight promptable architecture for medical image segmentation. SAMU builds upon medical SAM models for 3D image segmentation while remaining accessible to users without powerful computational resources. One clinical scenario with clear potential applications for promptable segmentation is in the field of radiation oncology, where physician annotation of anatomical structures is a labor-intensive, but necessary task [3]. One-point prompting for 3D segmentation in this context would save human resources while enabling improved patient care. SAMU outperforms previous methods on multiple diverse 3D medical image datasets. In addition to reducing computational requirements by utilizing a convolution UNet backbone, SAMU more completely leverages single-point prompt information by incorporating prompt features at multiple spatial resolutions in the decoder branch of the network. This may provide a stronger supervision signal than the prompt attention method used in other SAM architectures.

One limitation of our work is an inability to fully replicate the large-scale training datasets used by SAM-Med3D as the full data has not been made publicly available at this time. However, for the three datasets studied in this work, the authors of SAM-Med3D and FastSAM3D have confirmed previously that the same data splits were used for their model training and evaluation, thereby allowing a fair comparison of model performance. Further, while improved when compared to other baselines, SAMU still does not achieve outstanding segmentation performance. This may be due to the great heterogeneity in modalities and structures present in the training and evaluation data. Nevertheless, SAMU's ability to produce reasonable segmentations on diverse CT and MRI images for a wide range of targets reflects a promising generalizability as an efficient medical imaging foundation model. Due to its relatively low computational cost, SAMU is particularly well-suited to future study in meta-learning [6] frameworks which might improve model generalization to the heterogenous datasets prevalent in the medical domain. We believe SAMU represents an important contribution to 3D medical imaging segmentation which may allow for improvements in clinical workflows.

References

1. Baid, U., et al.: The rsna-asnr-miccai brats 2021 benchmark on brain tumor segmentation and radiogenomic classification. arXiv preprint arXiv:2107.02314 (2021)
2. Bakas, S., et al.: Advancing the cancer genome atlas glioma MRI collections with expert segmentation labels and radiomic features. Sci. Data **4**(1), 1–13 (2017)
3. Cardenas, C.E., Yang, J., Anderson, B.M., Court, L.E., Brock, K.B.: Advances in auto-segmentation. In: Seminars in Radiation Oncology. vol. 29, pp. 185–197. Elsevier (2019)
4. Cheng, J., et al.: Sam-med2d. arXiv preprint arXiv:2308.16184 (2023)
5. Dosovitskiy, A., et al.: An image is worth 16x16 words: Transformers for image recognition at scale. arXiv preprint arXiv:2010.11929 (2020)

6. Finn, C., Abbeel, P., Levine, S.: Model-agnostic meta-learning for fast adaptation of deep networks. In: International Conference on Machine Learning, pp. 1126–1135. PMLR (2017)
7. He, K., Chen, X., Xie, S., Li, Y., Dollár, P., Girshick, R.: Masked autoencoders are scalable vision learners. In: Proceedings of the IEEE/CVF Conference on Computer Vision and Pattern Recognition, pp. 16000–16009 (2022)
8. Ji, Y., et al.: Amos: a large-scale abdominal multi-organ benchmark for versatile medical image segmentation. Adv. Neural. Inf. Process. Syst. **35**, 36722–36732 (2022)
9. Kirillov, A et al.: Segment anything. In: Proceedings of the IEEE/CVF International Conference on Computer Vision, pp. 4015–4026 (2023)
10. Ma, J., He, Y., Li, F., Han, L., You, C., Wang, B.: Segment anything in medical images. Nat. Commun. **15**(1), 654 (2024)
11. Mazurowski, M.A., Dong, H., Gu, H., Yang, J., Konz, N., Zhang, Y.: Segment anything model for medical image analysis: an experimental study. Med. Image Anal. **89**, 102918 (2023)
12. Menze, B.H., et al.: The multimodal brain tumor image segmentation benchmark (brats). IEEE Trans. Med. Imaging **34**(10), 1993–2024 (2014)
13. Shen, Y., et al.: Fastsam3d: An efficient segment anything model for 3d volumetric medical images. arXiv preprint arXiv:2403.09827 (2024)
14. Wang, H., et al.: Sam-med3d. arXiv preprint arXiv:2310.15161v2 (2024)
15. Wasserthal, J., et al.: Totalsegmentator: robust segmentation of 104 anatomic structures in CT images. Radiol. Artif. Intell. **5**(5) (2023)
16. Zhou, L., Liu, H., Bae, J., He, J., Samaras, D., Prasanna, P.: Self pre-training with masked autoencoders for medical image classification and segmentation. In: 2023 IEEE 20th International Symposium on Biomedical Imaging (ISBI), pp. 1–6. IEEE (2023)

Anatomical Embedding-Based Training Method for Medical Image Segmentation Foundation Models

Mingrui Zhuang[1] , Rui Xu[1], Qinhe Zhang[2], Ailian Liu[2], Xin Fan[1], and Hongkai Wang[1,3(✉)]

[1] Dalian University of Technology, Dalian, China
`wang.hongkai@dlut.edu.cn`
[2] The First Affiliated Hospital of Dalian Medical University, Dalian, China
[3] Liaoning Key Laboratory of Integrated Circuit and Biomedical Electronic System, Dalian, China

Abstract. Existing training methods for medical image foundation models primarily focus on tasks such as image restoration, overlooking the potential of harnessing the inherent anatomical knowledge of the human body. The discrepancy between the training tasks of foundation models and downstream tasks often necessitates model fine-tuning for each specific application. An insufficient scale of the downstream training set can lead to catastrophic forgetting of the foundational model. To address these issues, we propose a novel unsupervised training method for medical image foundation models. Our approach incorporates an anatomical embedding task, enabling the model to generate anatomically related embeddings for each voxel. To expedite the training and accommodate large-scale models, we employ the strategy of momentum contrast learning, which is further enhanced to adapt to the task of anatomical embedding. To improve the model's performance for specific targets, we introduce the region contrastive loss, utilizing a small set of segmentation labels (e.g., five samples) to identify the focused regions during training. In our experiments, we pre-train the foundation model using a dataset of 4000 unlabeled abdominal CT scans with the downstream task being the few-shot learning of 13 abdominal organ segmentation. The results showed significant improvements in the downstream segmentation task, particularly in the scenarios with limited segmentation annotations, compared to methods without pre-training and similar foundation models. The trained models and the downstream training code have been open sourced at https://github.com/DlutMedimgGroup/Anatomy-Embedding-Foundation-Model.

Keywords: Foundation Model · Pre-Trained Model · Anatomical Embedding Learning

1 Introduction

Foundation models are commonly pretrained on extensive datasets to reduce the reliance on training data scale for specific downstream tasks. However, creating comprehensive medical image datasets for foundation training faces unique challenges, primarily due to

Z. Deng et al. (Eds.): MedAGI 2024, LNCS 15184, pp. 143–152, 2025.
https://doi.org/10.1007/978-3-031-73471-7_15

patient privacy protection restrictions and the demanding workload of data annotation. Several methods have emerged to transfer models pretrained on large-scale natural image datasets to the domain of medical imaging [8, 15, 18]. Despite the potential performance enhancements achieved by such methods, they lack training on native medical data and fail to utilize the inherent anatomical knowledge present in the data.

Most existing foundation models employ the approach of fine-tuning the decoder when applied to downstream tasks [3, 7, 13]. In this process, fine-tuning the original parameters of the foundation model with a small training sample can potentially lead to forgetting the pretrained knowledge. Even with the commonly used encoder freezing strategy, fully retraining the decoder still faces overfitting and imperfect robustness issues when dealing with small training samples.

Knowledge of anatomical relationships is crucial for improving the performance of medical foundation models [16]. Most existing methods train foundational models using tasks that are unrelated to anatomical information, such as image restoration. Anatomical embedding involves generating a feature vector for each pixel in the image based solely on its anatomical position (e.g., lower liver, middle left of the kidney, etc.). The anatomical embedding task holds tremendous potential for training foundational models. The training of the anatomical embedding task employs unsupervised contrastive learning, taking advantage of its unsupervised nature to reduce the reliance on data annotations in medical image processing tasks [11, 17]. The encoded knowledge obtained from the trained foundation model can be directly used as inputs for downstream tasks, eliminating the need for fine-tuning the decoder for individual tasks.

This study introduces an unsupervised training method for medical segmentation foundation models. The method utilizes an unsupervised anatomical position encoding task, utilizing a substantial volume of unlabeled data for training. For specific downstream tasks, the features from the trained foundation model can be directly inherited by downstream models, eliminating the need for specific fine-tuning of the decoder. This strategy mitigates the performance degradation typically associated with fine-tuning on small datasets. Different foundation models can be specifically trained for mainstream imaging examinations (e.g., abdominal CT, brain MRI, etc.), thereby enhancing the performance of various downstream tasks tailored to the corresponding image types. Our method offers the following innovative contributions:

(a) By employing the anatomical embedding task, we train the medical image foundation model to effectively utilize the inherent anatomical knowledge of the human body. The anatomical embedding can be directly utilized for downstream tasks, thereby avoiding the performance loss associated with model fine-tuning.

(b) In order to enable the utilization of larger model sizes and higher image resolutions, we employed the momentum contrast learning process strategy.

(c) To enhance the discriminative capability of the model's anatomical embedding features for important soft tissue organs, a region contrastive loss (RCL) designed for momentum-contrastive learning is proposed.

2   Methods

The proposed anatomical embedding-based framework of the foundation model train-ing is shown in Fig. 1. Initially, contrastive learning with an anatomical encoding task is employed to train the foundation model M_F. $M_{F'}$ has the same structure and ini-tialization as M_F, and its parameters are updated from M_F in the form of momentum [6]. During training, two patches with overlapping regions are randomly selected, as indicated by the yellow boxes in Fig. 1. The networks accept the patch as the input and outputs the anatomical embedding maps, where each pixel is a 64-dimensional fea-ture vector. Finally, anatomical contrastive loss is used to supervise that the pixels with same anatomical positions in both patches (indicated by the red dots) have same feature embeddings, while different positions have distinct feature embeddings.

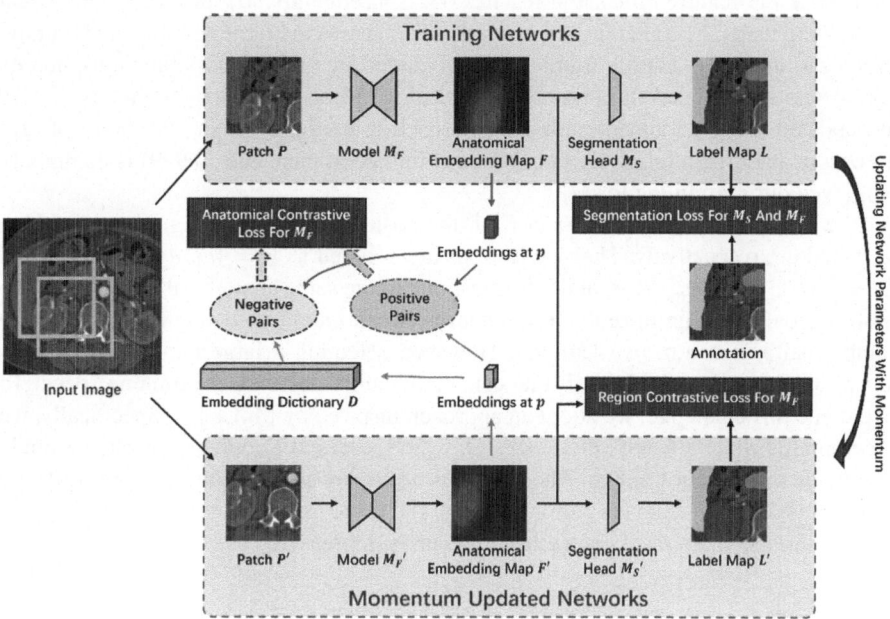

Fig. 1. Overview of the proposed framework.

To enhance the discriminative ability of the model towards important soft tissue organs with similar grayscale values, we propose a method using a limited number of labels (e.g., five samples) to guide the network's attention. Specifically, a lightweight fully connected segmentation head M_S is added after the foundation model. For the small amount of annotated data, a standard segmentation loss is employed to supervise the network. As the segmentation head is lightweight, this supervision primarily impacts the foundation model M_F. For the majority of unlabeled data, the result of momentum-updated segmentation head $M_{S'}$ is utilized for region contrastive loss. This loss function ensures that embeddings of the same region are close while the embeddings of different regions far apart.

2.1 Momentum Contrastive Learning for Large Training Dataset

When selecting patches with overlapping regions, we introduce randomness to enhance the diversity of the training data. The sizes of the two patches are determined randomly, with each side's length ranging from 0.9 to 1.1 times the preset value. The overlapping range in each direction is set as a fraction of the smaller side length between the two patches, varying from 0.5 to 1.0 times. Within the overlapping region, n_{pos} points are randomly selected as positive sample pairs for training, and this collection is denoted as $P = \{p_1, p_2, \ldots, p_{n_{pos}}\}$. Finally, both patches are resampled to the predetermined size and individually input into M_F and $M_F\prime$.

The foundation model, M_F, is based on a 3D U-Net architecture. In this study, we adopt a five-layer network as an example, with each layer comprising two residual units. Unlike a typical U-Net, the decoder part of M_F differs in that the dimensionality reduction stops once the feature dimension reaches 64. Consequently, M_F produces anatomical embedding maps, F, containing 64 channels. $M_F\prime$ shares the same structure and training initialization as M_F. During training, M_F is guided by multiple loss functions, and its parameters are updated through backpropagation. Meanwhile, the parameters of $M_F\prime$ are updated using a momentum-based approach to gradually align with those of M_F. To ensure stability, a relatively larger momentum coefficient (i.e., 0.999) is commonly employed during training for $M_F\prime$.

We extract features from F and $F\prime$ at the positions specified by p_i, denoting them as e_i and $e_i\prime$, respectively. Their collections are denoted as $E = \{e_1, e_2, \ldots, e_{n_{pos}}\}$ and $E\prime = \{e_1\prime, e_2\prime, \ldots, e_{n_{pos}}\prime\}$. e_i and $e_i\prime$ correspond to the same anatomical position p_i. The training goal is to minimize the differences between them by treating them as positive sample pairs in contrastive learning. However, selecting a large number of features as negative sample pairs in real-time can significantly decrease the training speed. To overcome this challenge, we adopt an approach inspired by MoCo [6]. Specifically, we incorporate $E\prime$ into a feature dictionary, D, which serves as a source of negative sample points for subsequent training. The positions of previously generated features in D are considered different from the current point. E and D are treated as negative sample pairs. As a result, the anatomical contrastive loss can be represented as

$$L_{AC} = -\log \frac{\exp\left(\sum_{i=1}^{n_{pos}} e_i \cdot e_i'/\tau\right)}{\exp\left(\sum_{i=1}^{n_{pos}} \sum_{j=1}^{n_{neg}} e_i \cdot d_j/\tau\right) + \exp\left(\sum_{i=1}^{n_{pos}} e_i \cdot e_i'/\tau\right)} \tag{1}$$

where n_{neg} denotes the size of D and τ is a temperature hyper-parameter. Due to the normalization operation applied to the output of M_F, $\|e_i\| = 1$. As a result, $e_i \cdot e_i\prime$ can be directly interpreted as the cosine similarity between them.

2.2 Guiding Network Attention Using a Few Annotations

To enhance the discriminative ability of the model towards important soft tissue organs with similar grayscale values, we propose a region supervision loss to fine-tune the network for downstream tasks. In our approach, we introduce a lightweight segmentation head, denoted as M_S, which consists of two fully connected layers. M_S is appended to the end of M_F, facilitating the conversion of anatomical embedding maps F into segmentation results L. For the small amount of annotated data (i.e., five samples), a common segmentation loss function, denoted as L_S, is employed. Notably, due to the lightweight nature of M_S, L_S significantly contributes to supervising M_F.

Similar to $M_{F'}$, $M_{S'}$ is a momentum-updated network based on M_S. Due to the larger momentum coefficient, the output of $M_{S'}$, represented by L', exhibits greater stability during the training process compared to L. L' is employed to supervise the network M_F in distinguishing the feature outputs between regions of interest. Inspired by the local contrastive loss [2], we propose a region contrastive loss designed for momentum-contrastive learning to supervise the similarity of features within the same region in F, while encouraging larger dissimilarity between features from different regions. This loss function, denoted as L_{RC}, can be represented as

$$L_{RC} = -\log \frac{1 + \exp\left(\sum_{m=1}^{C} \overline{e_m} \cdot \overline{e'_m}\right)}{1 + \exp\left(\sum_{m=1}^{C} \overline{e_m} \cdot \overline{e'_m}\right) + \exp\left(\sum_{m=1}^{C} \sum_{n=1, n \neq m}^{C} \overline{e_m} \cdot \overline{e'_n}\right)} \tag{2}$$

where C represents the number of segmentation categories. $\overline{e_m}$ and $\overline{e_{m'}}$ denote the averages of embeddings belonging to segmentation label m in the map E and E', respectively.

3   Experiments and Results

Datasets. To ensure the training effectiveness of the foundation model, we utilized the FLARE23 dataset [9], which comprises more than 4250 abdominal CT images, of which 4000 randomly selected data samples were used for unsupervised training of the foundation mode. To evaluate the performance of the foundation model, we employed the rest 250 samples with 13 abdominal organ labels to train a downstream segmentation network. Among these, 200 samples were allocated for training, while the remaining 50 samples were set aside for testing. Prior to training, all data underwent standardized preprocessing steps, including gray-scale clipping within the range of $[-500, 750]$, gray-scale normalization, and resampling to an image resolution of $1 \times 1 \times 1$ mm^3.

Implementation and Evaluation Criteria. The proposed framework is implemented on MONAI [1] and PyTorch. To accommodate the large model scale, a deep learning server equipped with 8 NVIDIA A800 GPUs was employed for training the foundation model. We empirically set $n_{pos} = 1000$ and $n_{neg} = 100000$ to strike a balance between training speed and resource utilization. During training, patch pairs with overlapping regions are alternately fed into M_F and $M_{F'}$, and the mean of the two losses is used as the final loss. In the first 500 epochs of training, only L_C is utilized, and in the subsequent 200 epochs, the segmentation head and the region loss function are introduced.

Foundation Model Test. To validate if the foundation model has learned the anatomical embedding from the training images, a test of inter-subject anatomical correspondence was conducted, as shown in Fig. 2. The template image is fed into the network to obtain anatomical embeddings for specific points. The predicted position of the prediction image is determined by selecting the anatomical embedding point with the highest cosine similarity to the input point. The examples shown in Fig. 2 represent scenarios where the target points are located both on the boundaries and within the organs. Notably, despite the anatomical structural differences between the template and predicted images, the network of this method succeeds in accurately identifying the correct anatomical correspondences. The improved correspondence accuracy can also be attributed to the introduction of momentum contrastive learning, enabling training of the foundation model at a spatial resolution of $1 \times 1 \times 1$ mm^3.

It is worth mentioning that the proposed training method may possess potential advantages in the context of swarm learning [12, 14]. According to formula (1), the computation of L_C depends solely on e_i, $e_{i'}$, and D, with D derived from $e_{i'}$. By sharing e_i and $e_{i'}$ among the swarm learning clients, it becomes feasible to perform synchronous training of the model across different centers under the same initialization conditions. By eliminating the need to transmit model parameters, the risk of model leakage is reduced. To guarantee data security, the transmitted anatomical embeddings are randomly sampled from discrete positions in the images and then shuffled, making it impossible to reconstruct the original data. This approach enables each client to leverage the entire dataset while maintaining data security. This eliminates the need for model aggregation algorithms (e.g., FedAvg [10]) and the associated performance losses.

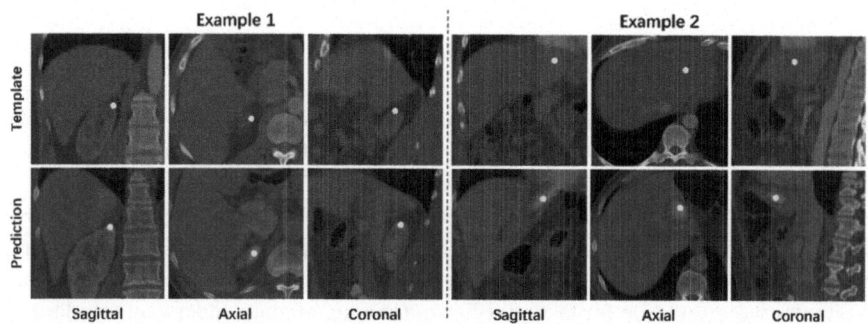

Fig. 2. Point-wise anatomical correspondence test results of the foundation model. The yellow dots indicate the input and output positions. The heatmap in the prediction image represents the similarity of features to the input points.

Downstream Segmentation Task. To evaluate the effectiveness of the trained foundation model in downstream tasks, we conducted validation using a segmentation task focused on 13 major abdominal organs. The downstream segmentation network is based on a conventional U-Net architecture, where the anatomical embeddings generated by the trained foundation model are combined with the original image as input. As control methods, we also employed the U-Net and the UNETR [5] without anatomical embedding input. To further validate the performance of the foundation model, we employed the Swin UNETR method [4], initialized with a self-supervised pre-trained Swin Transformer backbone [13], as an additional control. Furthermore, we conducted an ablation experiment where we excluded the proposed region contrast loss (RCL).

Figure 3 showcases a qualitative comparison of the segmentation results obtained from three pre-trained methods when trained with 20 data examples. The results demonstrate that our method exhibits the least number of flaws in the segmentation results. Figure 4 presents a comparison of the results obtained from each method when trained with 20 data examples. An additional set of 50 data samples was used as the test dataset. The networks utilizing anatomical embedding input outperformed non-pretraining methods across all segmentation targets. Notably, our method outperforms the pre-trained Swin UNETR in terms of segmentation results for the majority of targets. Furthermore, the incorporation of GLC (Global Local Contrast) further enhanced the segmentation accuracy for the majority of targets, confirming its effectiveness.

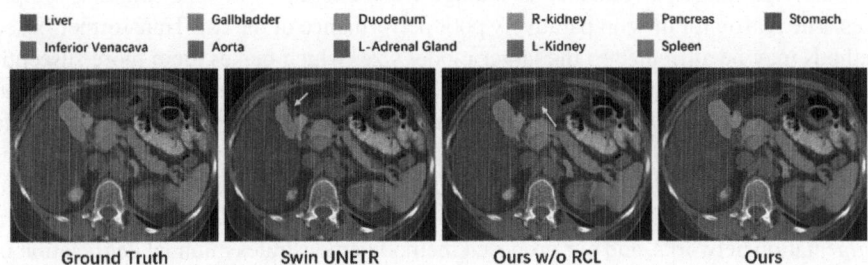

Fig. 3. Qualitative demonstration of segmentation results using different methods. The yellow arrows indicate noticeable defects.

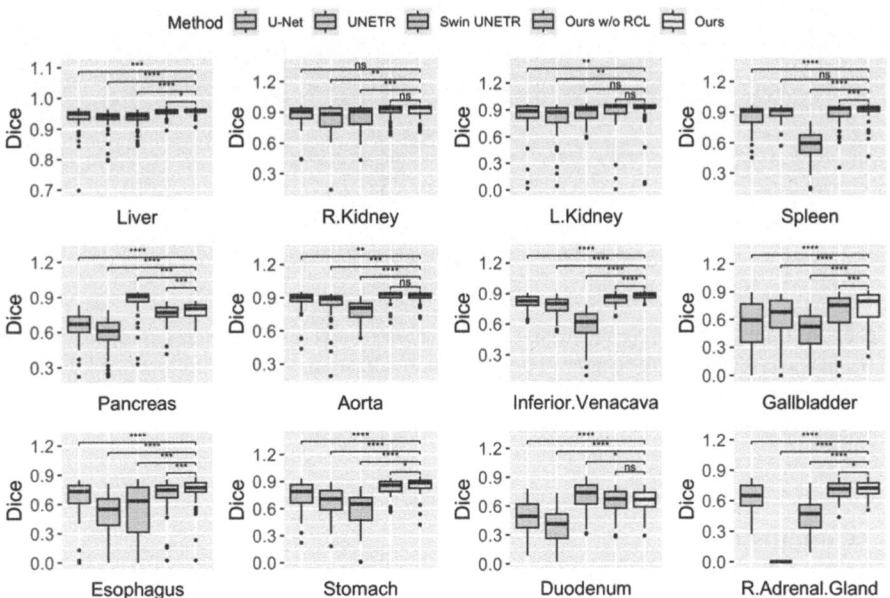

Fig. 4. Quantitative comparisons of segmentation performance using a training dataset consisting of 20 volumes. * indicates $p < .05$; ** indicates $p < .01$; *** indicates $p < .001$; **** indicates $p < .001$; **** indicates $p < .0001$.

Comparison with Existing Foundation Models. To compare our method with existing medical foundation models, we conducted a comprehensive comparison with Med-SAM[8]. MedSAM is officially pre-trained on the Flare22 dataset (similar to the Flare23 dataset we used) using much more computing resources (20 A100 GPUs), requiring an additional text prompt indicating the presence of each organ in each 2D slice. Our model yielded higher Dice score (0.84 ± 0.09) than MedSAM (0.74 ± 0.17).

The Robustness of Downstream Tasks. The foundation model also exhibits superior robustness than the compared method against small training sets and variations in input data grayscale. Figure 5a presents the performance of each method under different training set sizes. The results reveal that methods pre-trained with the foundation model exhibit a significant performance advantage when dealing with small training sample sizes. The reason for the comparatively poor performance of the two Transformer-based methods may be attributed to the larger model size, which makes them more susceptible to the limited training dataset. Figure 5b compares the robustness of each method to grayscale transformations. By default, the range for grayscale clipping and grayscale normalization is $[-500, 750]$. During grayscale transformation, this range is proportionally scaled down. The degree of range transformation is depicted on the horizontal axis of Fig. 5b. The results indicate that pre-trained models exhibit robustness for downstream segmentation networks, and our proposed method demonstrates minimal degradation in performance when the transformation degree is kept below 48%.

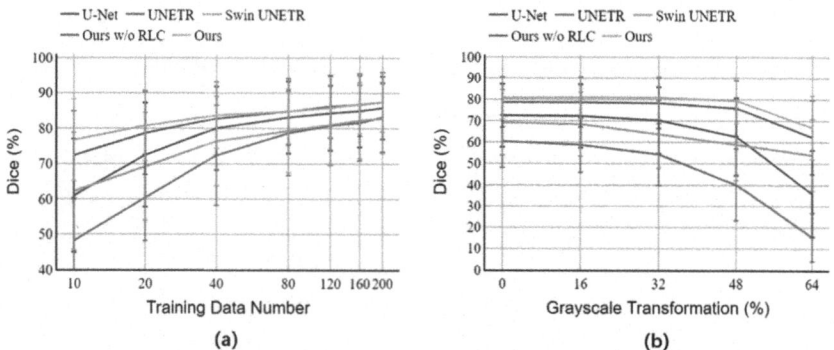

Fig. 5. (a) Impact of training set size on segmentation results. (b) Robustness comparison. The horizontal axis represents the degree of grayscale normalization range transformation of the input data.

4   Conclusion

In this paper, we propose a novel training framework for medical image foundation models. Our approach leverages the anatomical embedding task to train the foundational model, eliminating the need for downstream fine-tuning and mitigating the associated performance degradation. By incorporating the proposed momentum contrastive learning method and region contrastive loss, we have successfully improved training efficiency and enhanced the model's ability to discern soft tissues. Through experiments, we demonstrate the effectiveness of our approach through the foundation model and downstream segmentation model results.

Acknowledgments. This work was supported in part by the National Key Research and Development Program No. 2020YFB1711500, 2020YFB1711501 and 2020YFB1711503, the general program of National Natural Science Fund of China (No. 81971693, 61971445), the funding of Dalian Key Laboratory of Digital Medicine for Critical Diseases, the Fundamental Research Funds for the Central Universities (No. DUT22YG229 and DUT22YG205), the funding of Liaoning Key Lab of IC & BME System and Dalian Engineering Research Center for Artificial Intelligence in Medical Imaging.

Disclosure of Interests. The authors have no competing interests to declare that are relevant to the content of this article.

References

1. Cardoso, M.J., et al.: MONAI: an open-source framework for deep learning in healthcare, http://arxiv.org/abs/2211.02701 (2022). https://doi.org/10.48550/arXiv.2211.02701
2. Chaitanya, K., et al.: Local contrastive loss with pseudo-label based self-training for semi-supervised medical image segmentation. Med. Image Anal. **87**, 102792 (2023). https://doi.org/10.1016/j.media.2023.102792
3. Dosovitskiy, A., et al.: An Image is Worth 16×16 Words: Transformers for Image Recognition at Scale. http://arxiv.org/abs/2010.11929, (2021). https://doi.org/10.48550/arXiv.2010.11929
4. Hatamizadeh, A., et al.: Swin UNETR: swin transformers for semantic segmentation of brain tumors in MRI images. In: Crimi, A., Bakas, S. (eds.) Brainlesion: Glioma, Multiple Sclerosis, Stroke and Traumatic Brain Injuries. pp. 272–284. Springer, Cham (2022). https://doi.org/10.1007/978-3-031-08999-2_22
5. Hatamizadeh, A., et al.: UNETR: transformers for 3D Medical Image Segmentation. http://arxiv.org/abs/2103.10504 (2021). https://doi.org/10.48550/arXiv.2103.10504
6. He, K., et al.: Momentum contrast for unsupervised visual representation learning. In: 2020 IEEE/CVF Conference on Computer Vision and Pattern Recognition (CVPR), pp. 9726–9735. IEEE, Seattle, WA, USA (2020). https://doi.org/10.1109/CVPR42600.2020.00975
7. Liu, Z., et al.: Swin transformer: hierarchical vision transformer using shifted windows. Presented at the Proceedings of the IEEE/CVF International Conference on Computer Vision (2021)
8. Ma, J., et al.: Segment anything in medical images. Nat Commun. **15**, 1, 654 (2024). https://doi.org/10.1038/s41467-024-44824-z
9. Ma, J., et al.: Unleashing the strengths of unlabeled data in pan-cancer abdominal organ quantification: the FLARE22 challenge. http://arxiv.org/abs/2308.05862 (2023). https://doi.org/10.48550/arXiv.2308.05862
10. McMahan, B., et al.: Communication-efficient learning of deep networks from decentralized data. In: Proceedings of the 20th International Conference on Artificial Intelligence and Statistics, pp. 1273–1282. PMLR (2017)
11. Park, T., et al.: Contrastive learning for unpaired image-to-image translation. http://arxiv.org/abs/2007.15651 (2020). https://doi.org/10.48550/arXiv.2007.15651

12. Saldanha, O.L., et al.: Swarm learning for decentralized artificial intelligence in cancer histopathology. Nat. Med. **28**(6), 1232–1239 (2022). https://doi.org/10.1038/s41591-022-017 68-5
13. Tang, Y., et al.: Self-supervised pre-training of swin transformers for 3D medical image analysis. Presented at the Proceedings of the IEEE/CVF Conference on Computer Vision and Pattern Recognition (2022)
14. Warnat-Herresthal, S., et al.: Swarm Learning for decentralized and confidential clinical machine learning. Nature **594**(7862), 265–270 (2021). https://doi.org/10.1038/s41586-021-03583-3
15. Wu, J., et al.: Medical SAM adapter: adapting segment anything model for medical image segmentation. http://arxiv.org/abs/2304.12620 (2023)
16. Yan, K., et al.: SAM: self-supervised learning of pixel-wise anatomical embeddings in radiological images. IEEE Trans. Med. Imaging **41**(10), 2658–2669 (2022). https://doi.org/10.1109/TMI.2022.3169003
17. Yu, Z., et al.: Cross-grained contrastive representation for unsupervised lesion segmentation in medical images. In: 2023 IEEE/CVF International Conference on Computer Vision Workshops (ICCVW), pp. 2339–2346 (2023). https://doi.org/10.1109/ICCVW60793.2023.00248
18. Zhang, Y., et al.: Input augmentation with SAM: boosting medical image segmentation with segmentation foundation model. In: Celebi, M.E., et al. (eds.) Medical Image Computing and Computer Assisted Intervention – MICCAI 2023 Workshops. pp. 129–139. Springer Nature Switzerland, Cham (2023). https://doi.org/10.1007/978-3-031-47401-9_13

Boosting Vision-Language Models for Histopathology Classification: Predict All at Once

Maxime Zanella[1,2]([✉]) [iD], Fereshteh Shakeri[3,4] [iD], Yunshi Huang[3,4] [iD],
Houda Bahig[4] [iD], and Ismail Ben Ayed[3,4] [iD]

[1] Université Catholique de Louvain (UCLouvain), Louvain-La-Neuve, Belgium
maxime.zanella@uclouvain.be
[2] Université de Mons (UMons), Mons, Belgium
[3] École de Technologie Supérieure (ÉTS), Montréal, Canada
fereshteh.shakeri.1@etsmtl.net
[4] Centre de Recherche du Centre Hospitalier de l'Université de Montréal
(CRCHUM), Montréal, Canada

Abstract. The development of vision-language models (VLMs) for histo-pathology has shown promising new usages and zero-shot performances. However, current approaches, which decompose large slides into smaller patches, focus solely on inductive classification, *i.e.*, prediction for each patch is made independently of the other patches in the target test data. We extend the capability of these large models by introducing a transductive approach. By using text-based predictions and affinity relationships among patches, our approach leverages the strong zero-shot capabilities of these new VLMs without any additional labels. Our experiments cover four histopathology datasets and five different VLMs. Operating solely in the embedding space (*i.e.*, in a black-box setting), our approach is highly efficient, processing 10^5 patches in just a few seconds, and shows significant accuracy improvements over inductive zero-shot classification. Code available at https://github.com/FereshteShakeri/Histo-TransCLIP.

Keywords: Histopathology · Medical VLMs · Zero-Shot Learning · Transductive Inference · Efficient Adaptation

1 Introduction

Histology slides obtained from Whole Slide Image (WSI) [18] scanners play a crucial role in cancer diagnosis and staging [16]. These slides offer a detailed view of diseased tissues, aiding in the determination of treatment options. Pathologists primarily diagnose cancers by examining WSIs to identify different tissue types. However, manually analyzing these WSIs imposes a significant workload, leading to substantial delays in reporting time. Moreover, in real clinical environments, the classification of cancer-related tissues is highly diverse, encompassing various

---

M. Zanella and F. Shakeri—are Equally Contribution.

© The Author(s), under exclusive license to Springer Nature Switzerland AG 2025
Z. Deng et al. (Eds.): MedAGI 2024, LNCS 15184, pp. 153–162, 2025.
https://doi.org/10.1007/978-3-031-73471-7_16

cancer sites. Even within a single cancer site, tasks can vary in their levels of class granularity. Therefore, automating tissue-type classification in histology images ([1,12,19,20,25] to list a few) holds significant clinical value but is hindered by the difficulty of collecting large labeled datasets and the variability of fine-grained labels.

The advent of multi-modal learning methods that process and integrate information from diverse modalities has alleviated some issues of training fine-grained classifiers and collecting costly labeled data. In particular, vision-language models (VLMs) such as CLIP [21] and ALIGN [9] have gained popularity in computer vision, and demonstrated promising generalization capabilities across various downstream tasks. These so-called foundation models jointly train vision and text embeddings using contrastive learning on large-scale image-text datasets. This new multi-modal paradigm can naturally be extended to clinical scenarios, where combinations of multiple data modalities–mainly texts and images–are often adopted to obtain more accurate and comprehensive diagnosis. For example, clinical notes and pathology reports, alongside histopathology slides, are commonly used for throughout analysis [6]. However, the direct application of deep learning techniques, more specifically vision-language pre-training strategies, to medical imaging is complex, due to the lack of fine-grained expert medical knowledge, which is required to capture specialized information [4]. This issue has been partly addressed for histopathology slides by collecting diverse data from scientific publications, Twitter, or even YouTube videos [7,8,15].

Current usage of such models predominantly align with the *inductive* paradigm, i.e., inference for each test sample is performed independently from the other samples within the test dataset. In contrast, *transduction* performs joint inference on all the test samples of a task, leveraging the statistics of the target unlabeled data [10,26]. Transduction has primarily been explored for few-shot classification of natural images, to tackle the inherent challenges of training under limited supervision [3,5]. These techniques utilize labeled samples to transfer information to unlabeled test data. Interestingly, in the novel multi-modal paradigm introduced by VLMs, supervision can be instead provided through textual descriptions of each classes (prompts), in a zero-shot setting, *e.g.,* a pathology tissue showing [class name]. Along with their corresponding representation derived from the language encoder, similarities between text and image embeddings can be leveraged to enable transductive inference even in the *zero-shot* scenario, as pointed by recent works in computer vision [17,29] (see Fig. 1).

Contributions. With the ongoing development of foundation models in medical imaging and specifically histopathology, and the potential application of transductive inference, our objective is to improve zero-shot predictions of VLMs within this framework. Our main contributions can be summarized as follows:

- We compare the zero-shot performance of vision-language models for histology and propose an effective transductive method to significantly boost their accuracy by leveraging the structure among patches during inference.
- Our transductive approach does not require labels; instead, it utilizes text-based predictions as regularization.
- To alleviate the computational workload, our method relies on the pre-computed features only, without access to the pre-trained weights, thus accommodating black-box constraints. This makes it feasible to process very large-scale slides in a matter of seconds.

2   Related Work

VLMs for Histology. Unlike natural images, which are often available in millions (*e.g.,* CLIP [21] is trained on 400 M image-text pairs), clinical image-text pairs are more challenging to amass. Similar to other works introducing VLMs for medical imaging (*e.g.,* for radiology [27,28,30], or ophthalmology [23]), several VLMs for computational pathology has appeared recently, differentiating themselves primarily through their data collection and curation methodologies. PLIP [7] curates OpenPath, a large dataset of pathology images paired with text descriptions. Quilt-1M [8] stands as one of the largest vision-language histopathology dataset to date, comprising 1 million image/text pairs sourced from YouTube videos. More recently CONCH [15] integrates parts of the PubMed Central Open Access Dataset yielding 1.17 million samples. As these new VLMs have been developed in a short amount of time, determining the most suitable one is not straightforward. Therefore, we provide a comparison of these models and demonstrate the applicability of our approach across each of them.

Transductive Learning. In the few-shot literature solely based on vision models, transduction leverages both the few labeled samples and unlabeled test data [10,26], outperforming inductive methods [3,5,14,31]. This setting, widely explored in computer vision, has been recently deployed in histopathology [22], using the annotations of a few patches from slides of liver. However, previously mentioned transductive methods have been shown to suffer from significant performance drops when applied to VLMs [17,29]. This motivated a few, very recent transductive methods in computer vision, focusing on natural images and explicitly leveraging the textual modality along the image embeddings [17,29]. In contrast to [22], our work exploits the findings and transductive-inference zero-shot objective in [29], aiming to boost the predictive accuracy of pretrained histopathology VLMs without any supervision.

3   Method

In this section, we describe the Histo-TransCLIP objective function for transductive inference in VLMs, for the K-class prediction problem. This objective function depends on two types of variables: (i) assignment variables $\mathbf{z}_i =$

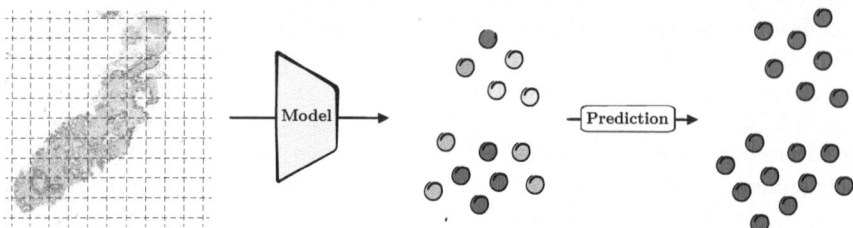

(a) In the typical inductive setting, a model is trained and then used to infer on each patch separately. This approach can be efficient when large annotated datasets for each task are available. This procedure often involves predicting the most probable class (argmax).

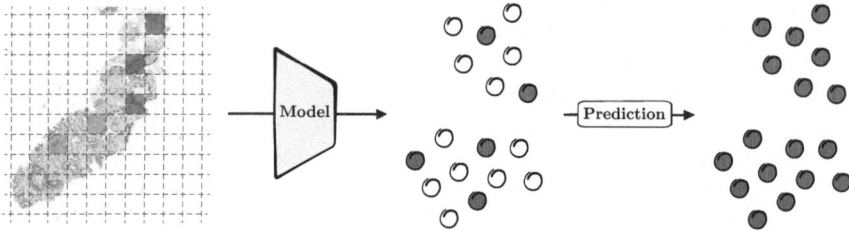

(b) In the traditional transductive few-shot setting, a pre-trained encoder (e.g., on ImageNet or large-scale histology dataset) requires manual annotations for the new task to propagate information from labeled to unlabeled samples. This process often involves measuring affinities or distances between encoded samples.

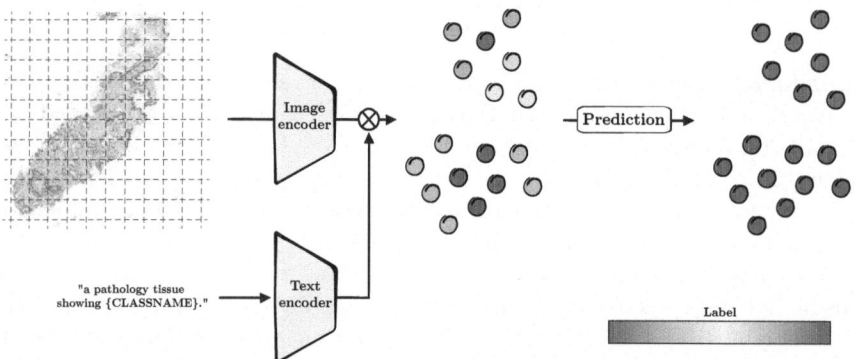

(c) VLMs leverage textual descriptions of each class to generate pseudo-labels without any manual annotation. These initial predictions can then be refined, for example, by leveraging the data structure.

Fig. 1. Illustration depicting histopathology classification in the inductive setting (a), the commonly-used few-shot transductive setting (b), and the zero-shot transductive setting enabled by VLMs (c).

$(z_{i,k})_{1 \leq k \leq K} \in \Delta_K$, for each patch $i \in \mathcal{Q}$; and (ii) Gaussian Mixture Model (GMM) parameters $\boldsymbol{\mu} = (\boldsymbol{\mu}_k)_{1 \leq k \leq K}$ and $\boldsymbol{\Sigma}$. We will first detail the main components of Histo-TransCLIP, before deriving the overall procedure.

Gaussian Modelization. We modelize the likelihood of the target data as a balanced mixture of multivariate Gaussian distributions, each representing a class k, parameterized by mean vector $\boldsymbol{\mu}_k$ and a diagonal (shared among classes) covariance matrix $\boldsymbol{\Sigma}$:

$$p_{i,k} \propto \det(\boldsymbol{\Sigma})^{-\frac{1}{2}} \exp\left(-\frac{1}{2}(\mathbf{f}_i - \boldsymbol{\mu}_k)^\top \boldsymbol{\Sigma}^{-1}(\mathbf{f}_i - \boldsymbol{\mu}_k)\right)$$

where $\mathbf{f}_i$ represents the encoding of patch i.

Text-based Predictions. When dealing with a zero-shot classification problem based on a VLM, and given a set of K candidate classes, one can get textual embeddings $\mathbf{t}_k$ (*e.g.*, from a `pathology tissue showing [kth class name]`, $k = 1, \ldots, K$). Then pseudo-labels can be obtained by evaluating the softmax function of the cosine similarities between these two encoded modalities with τ being a temperature parameter:

$$\hat{y}_{i,k} = \frac{\exp(\tau \mathbf{f}_i^\top \mathbf{t}_k)}{\sum_j \exp(\tau \mathbf{f}_i^\top \mathbf{t}_j)} \tag{1}$$

Laplacian Regularization. Laplacian regularizers are widely used in the context of graph/spectral clustering. This term encourages related samples (*i.e.*, pairs of patches with high affinity $w_{i,j}$) to have similar label assignments. We build affinities based on the cosine similarities of each patch representation:

$$w_{ij} = \mathbf{f}_i^\top \mathbf{f}_j \tag{2}$$

In fact, affinity relations can be modified for each specific use-case, allowing to inject knowledge in the optimization process. In our case, we can leverage the strong embedding capabilities of the image encoder to regularize the transductive procedure. In practice, to reduce memory needs, we sparsify the matrix by retaining only the 3 nearest neighbors of each patch.

Objective Function. We minimize the following objective:

$$\mathcal{L}(\mathbf{z}, \boldsymbol{\mu}, \boldsymbol{\Sigma}) = \underbrace{-\frac{1}{|\mathcal{Q}|} \sum_{i \in \mathcal{Q}} \mathbf{z}_i^\top \log(\mathbf{p}_i)}_{\text{GMM clustering}} - \underbrace{\sum_{i \in \mathcal{Q}} \sum_{j \in \mathcal{Q}} w_{ij} \mathbf{z}_i^\top \mathbf{z}_j}_{\text{Laplacian regularization}} + \underbrace{\sum_{i \in \mathcal{Q}} \mathrm{KL}(\mathbf{z}_i \| \hat{\mathbf{y}}_i)}_{\text{Prediction penalty}} \tag{3}$$

The Kullback-Leibler (KL) term encourages the prediction $\mathbf{z}_i$ not to deviate significantly from the zero-shot prediction $\hat{\mathbf{y}}_i$, thereby providing text supervision without the need of any labels.

Procedure. We refer to [29] for the technical details about the derivation. Optimizing (3), subject to simplex constraints, we obtain the following decoupled

---

Pseudocode 1: Histo-TransCLIP procedure for transductive inference alternates between assignments and GMM-parameters updates.

Input: $\mathbf{f}$ are the image embeddings, $\mathbf{t}$ are the text/class embeddings, τ is the temperature scaling used during each VLM pretraining.

---

1 **function** Histo-TransCLIP($\mathbf{f}$, $\mathbf{t}$, τ)

   // Text-based pseudo-labels $\hat{y}$

2   $\hat{\mathbf{y}}_i = \mathrm{softmax}(\tau \mathbf{f}_i^T \mathbf{t})$ $\forall i$

   // Initialize z, μ, Σ

3   $\mathbf{z}_i = \hat{\mathbf{y}}_i$ $\forall i$

4   $\boldsymbol{\mu}_k = \mathrm{top\_confident\_average}(\mathbf{f}, \hat{\mathbf{y}})$ $\forall k$

5   $\mathrm{diag}(\boldsymbol{\Sigma}) = \frac{1}{n\_features}$

   // Iterative procedure

6   **while** not_converged **do**

7   **for** l = 1:... **do**

8         $\mathbf{z}_i^{(l+1)} = \dfrac{\hat{\mathbf{y}}_i \odot \exp(\log(\mathbf{p}_i) + \sum_{j \in \mathcal{Q}} w_{ij} \mathbf{z}_j^{(l)})}{(\hat{\mathbf{y}}_i \odot \exp(\log(\mathbf{p}_i) + \sum_{j \in \mathcal{Q}} w_{ij} \mathbf{z}_j^{(l)}))^\top \mathbb{1}_K}$ $\forall i$

9      $\boldsymbol{\mu}_k = \dfrac{\sum_{i \in \mathcal{Q}} z_{i,k} \mathbf{f}_i}{\sum_{i \in \mathcal{Q}} z_{i,k}}$ $\forall k$

10      $\mathrm{diag}(\boldsymbol{\Sigma}) = \frac{1}{|\mathcal{Q}|} \sum_{i \in \mathcal{Q}} \sum_k z_{i,k} (\mathbf{f}_i - \boldsymbol{\mu}_k)^2$

11   **return** z

---

update rules for the assignment variables, which can be computed in parallel for all samples (*i.e.*, patches) at a given iteration l:

$$\mathbf{z}_i^{(l+1)} = \frac{\hat{\mathbf{y}}_i \odot \exp(\log(\mathbf{p}_i) + \sum_{j \in \mathcal{D}} w_{ij} \mathbf{z}_j^{(l)})}{(\hat{\mathbf{y}}_i \odot \exp(\log(\mathbf{p}_i) + \sum_{j \in \mathcal{D}} w_{ij} \mathbf{z}_j^{(l)}))^\top \mathbb{1}_K} \tag{4}$$

Note how each assignment $\mathbf{z}_i$ depends on its neighbors. This update must be computed iteratively until convergence, enabling assignments to propagate from the GMM likelihood to neighboring samples, weighted by their affinity. Since these updates are decoupled, this step can be parallelized efficiently (see runtime in Table 2). With other variables fixed, we then have the following closed-form updates for the GMM parameters:

$$\boldsymbol{\mu}_k = \frac{\sum_{i \in \mathcal{Q}} z_{i,k} \mathbf{f}_i}{\sum_{i \in \mathcal{Q}} z_{i,k}} \tag{5}$$

$$\mathrm{diag}(\boldsymbol{\Sigma}) = \frac{1}{|\mathcal{Q}|} \sum_{i \in \mathcal{Q}} \sum_k z_{i,k} (\mathbf{f}_i - \boldsymbol{\mu}_k)^2 \tag{6}$$

The procedure is summarized in Pseudocode 1 and alternates between solving (4) to get assignments for each patch and computing the GMM parameters (5) (6) according to those assignments until convergence (see proof in [29]).

Table 1. Zero-shot and Histo-TransCLIP performance on top of various VLMs. Best values are highlighted in **bold**. $\Delta_{\text{transductive}}$ is the average accuracy gain brought by our transductive approach.

Dataset	Method	Model				
		CLIP	Quilt-B16	Quilt-B32	PLIP	CONCH
SICAP-MIL	Zero-shot	**29.85**	40.44	**35.04**	46.84	27.71
	Histo-TransCLIP	24.72	**58.49**	28.18	**53.23**	**32.58**
LC(Lung)	Zero-shot	**31.46**	43.00	76.24	84.96	84.81
	Histo-TransCLIP	25.62	**50.53**	**93.93**	**93.80**	**96.29**
SKINCANCER	Zero-shot	4.20	15.38	39.71	22.90	58.53
	Histo-TransCLIP	**11.46**	**33.33**	**48.80**	**36.72**	**66.22**
NCT-CRC	Zero-shot	25.39	29.61	53.73	63.17	66.27
	Histo-TransCLIP	**39.61**	**48.40**	**58.13**	**77.53**	**70.36**
Average	Zero-shot	22.73	32.1	51.18	54.47	59.33
	Histo-TransCLIP	**25.35**	**47.69**	**57.26**	**65.32**	**66.36**
	$\Delta_{\text{transductive}}$	**+2.62**	**+15.59**	**+6.08**	**+10.85**	**+7.03**

4   Experiments

We conduct a comprehensive comparison of several vision-language models pretrained on histology images, namely PLIP [7], QUILT [8] (for which we report two versions) and CONCH [15]. Text embeddings for each category are obtained following the specific 22 prompts used for CONCH (only one name is assigned to each target class), which are then averaged to get a single textual embedding per class. Numerical results are top-1 accuracy which compare zero-shot prediction (*i.e.*, inductive inference) and Histo-TransCLIP (*i.e.*, transductive inference).

Datasets. We study different histopathology classification tasks on various organs and cancer types [2,11,13,24]. Specifically, *NCT-CRC* [11] comprises patches of colorectal adenocarcinoma categorized into 9 classes, *SICAP-MIL* [24] includes 4 prostate cancer grading, *SKINCANCER* [13] is annotated with 9 anatomical tissue structures, and *LC25000(Lung)* [2] focuses on 3 classes of lung cancer. These diverse benchmarks enable the study of the generalization capability of VLMs pretrained on histology images and provide a thorough assessment of our transductive approach.

Results. Table 1 presents a comparative analysis of zero-shot performance and the improvement achieved by Histo-TransCLIP. The lower classification accuracy of CLIP emphasizes the need for VLMs specifically pretrained on histology. Notably, the recently proposed CONCH model demonstrates the highest average accuracy. Note that the variation in zero-shot accuracies compared to the original paper values is largely influenced by the choice of prompt templates,

Table 2. Features denotes the runtime to pre-compute the image and text embeddings, Histo-TransCLIP denotes the runtime of our transductive procedure once embeddings are provided. Experiments were conducted on a single NVIDIA GeForce RTX 3090 (24 Gb) GPU.

#Patches	Features	Histo-TransCLIP
10^2	~1 sec.	~0.1 sec.
10^3	~4 sec.	~0.2 sec.
10^4	~28 sec.	~0.4 sec.
10^5	~5 min.	~6 sec.

for instance PLIP zero-shot results are significantly improved. This yields interesting questions on prompt sensitivy as discussed for future work in Sect. 5. Histo-TransCLIP consistently enhances performance significantly, highlighting the benefits of its transductive approach. Only in a few cases does the accuracy of Histo-TransCLIP drop, particularly when zero-shot performance is low due to direct regularization with initial text predictions. In most cases, Histo-TransCLIP effectively enhances performance, even on tasks initially achieving high accuracy, showcasing its strong ability to refine slightly misaligned text predictions for various VLMs.

Computational Workload. Table 2 details the computational overhead associated with Quilt-B16 visual and textual feature extraction, alongside the implementation of Histo-TransCLIP across varying patch numbers in the *NCT-CRC* dataset. While the time for feature extraction increases with the number of patches, the additional workload introduced by Histo-TransCLIP remains negligible. This shows transduction can importantly improve performance while maintaining black-box adaptation (*i.e.*, without accessing the model's parameters) and without adding any notable additional workload.

5   Conclusion

We have demonstrated the significant value that transduction can bring to histology. By leveraging text-based predictions through a Kullback-Leibler divergence penalty and incorporating shared information among patches with Laplacian regularization, our approach significantly enhances the performance of vision-language models. Notably, our method is highly efficient and does not require additional labels or access to model parameters.

Future Work. Our approach can be naturally extended to the few-shot setting. Additionally, the quality of the prompts, *i.e.*, the textual descriptions of each class, can significantly impact the final zero-shot performance. Studying this impact is undoubtedly valuable for safer applications. Finally, while our current work focuses on transduction using patches from multiple slides, a more constrained and valuable application would involve transduction on patches from a single slide to improve performance on a per-patient basis.

Acknowledgement. M. Zanella is funded by the Walloon region under grant No. 2010235 (ARIAC by DIGITALWALLONIA4.AI). F. Shakeri is funded by Natural Sciences and Engineering Research Council of Canada (NSERC) and Canadian Institutes of Health Research (CIHR).

Disclosure of Interests. The authors have no competing interests to declare that are relevant to the content of this article.

References

1. Bilgin, C., Demir, C., Nagi, C., Yener, B.: Cell-graph mining for breast tissue modeling and classification. In: 2007 29th Annual International Conference of the IEEE Engineering in Medicine and Biology Society, pp. 5311–5314. IEEE (2007)
2. Borkowski, A.A., Bui, M.M., Thomas, L.B., Wilson, C.P., DeLand, L.A., Mastorides, S.M.: Lung and colon cancer histopathological image dataset (lc25000). arXiv preprint arXiv:1912.12142 (2019)
3. Boudiaf, M., Ziko, I., Rony, J., Dolz, J., Piantanida, P., Ben Ayed, I.: Information maximization for few-shot learning. Adv. Neural. Inf. Process. Syst. **33**, 2445–2457 (2020)
4. Chen, X., et al.: Recent advances and clinical applications of deep learning in medical image analysis. Med. Image Anal. **79** (2022)
5. Dhillon, G.S., Chaudhari, P., Ravichandran, A., Soatto, S.: A baseline for few-shot image classification. In: International Conference on Learning Representations (2019)
6. Hartsock, I., Rasool, G.: Vision-language models for medical report generation and visual question answering: A review. CoRR **abs/2403.02469** (2024). https://doi.org/10.48550/ARXIV.2403.02469, https://doi.org/10.48550/arXiv.2403.02469
7. Huang, Z., Bianchi, F., Yuksekgonul, M., Montine, T., Zou, J.: A visual-language foundation model for pathology image analysis using medical twitter. Nat. Med. **29**, 1–10 (2023)
8. Ikezogwo, W.O., et al.: Quilt-1m: One million image-text pairs for histopathology. arXiv preprint arXiv:2306.11207 (2023)
9. Jia, C., et al.: Scaling up visual and vision-language representation learning with noisy text supervision. In: International Conference on Machine Learning, pp. 4904–4916 (2021)
10. Joachims, T.: Transductive inference for text classification using support vector machines. In: Proceedings of the Sixteenth International Conference on Machine Learning, pp. 200–209 (1999)
11. Kather, J.N., Halama, N., Marx, A.: 100,000 histological images of human colorectal cancer and healthy tissue. Zenodo10 **5281** (2018)
12. Komura, D., Ishikawa, S.: Machine learning methods for histopathological image analysis. Comput. Struct. Biotechnol. J. **16**, 34–42 (2018)
13. Kriegsmann, K., et al.: Deep learning for the detection of anatomical tissue structures and neoplasms of the skin on scanned histopathological tissue sections. Front. Oncol. **12**, 1022967 (2022)
14. Liu, J., Song, L., Qin, Y.: Prototype rectification for few-shot learning. In: Vedaldi, A., Bischof, H., Brox, T., Frahm, J.M., (eds.) Computer Vision – ECCV 2020. ECCV 2020. Lecture Notes in Computer Science(), vol. 12346, pp. 741–756. Springer, Cham (2020). https://doi.org/10.1007/978-3-030-58452-8_43

15. Lu, M.Y., et al.: A visual-language foundation model for computational pathology. Nat. Med. **30**, 863–874 (2024)
16. Madabhushi, A.: Digital pathology image analysis: opportunities and challenges. Imaging Med. **1**(1), 7 (2009)
17. Martin, S., Huang, Y., Shakeri, F., Pesquet, J.C., Ben Ayed, I.: Transductive zero-shot and few-shot clip. In: Proceedings of the IEEE/CVF Conference on Computer Vision and Pattern Recognition, pp. 28816–28826 (2024)
18. Pantanowitz, L.: Digital images and the future of digital pathology. J. Pathol. Inform. **1** (2010)
19. Petushi, S., Garcia, F.U., Haber, M.M., Katsinis, C., Tozeren, A.: Large-scale computations on histology images reveal grade-differentiating parameters for breast cancer. BMC Med. Imaging **6**(1), 1–11 (2006)
20. Qureshi, H., Sertel, O., Rajpoot, N., Wilson, R., Gurcan, M.: Adaptive discriminant wavelet packet transform and local binary patterns for meningioma subtype classification. In: Metaxas, D., Axel, L., Fichtinger, G., Székely, G. (eds.) Medical Image Computing and Computer-Assisted Intervention – MICCAI 2008. MICCAI 2008. Lecture Notes in Computer Science, vol. 5242, pp. 196–204. Springer, Berlin, Heidelberg (2008). https://doi.org/10.1007/978-3-540-85990-1_24
21. Radford, A., et al.: Learning transferable visual models from natural language supervision. In: International Conference on Machine Learning, pp. 8748–8763 (2021)
22. Sadraoui, A., et al.: A transductive few-shot learning approach for classification of digital histopathological slides from liver cancer. In: IEEE International Symposium on Biomedical Imaging (ISBI) (2024)
23. Silva-Rodriguez, J., Chakor, H., Kobbi, R., Dolz, J., Ayed, I.B.: A foundation language-image model of the retina (flair): Encoding expert knowledge in text supervision. ArXiv Preprint (2023)
24. Silva-Rodríguez, J., Schmidt, A., Sales, M.A., Molina, R., Naranjo, V.: Proportion constrained weakly supervised histopathology image classification. Comput. Biol. Med. **147**, 105714 (2022)
25. Tabesh, A., et al.: Multifeature prostate cancer diagnosis and gleason grading of histological images. IEEE Trans. Med. Imaging **26**(10), 1366–1378 (2007)
26. Vapnik, V.: An overview of statistical learning theory. IEEE Trans. Neural Netw. **10**(5), 988–999 (1999). https://doi.org/10.1109/72.788640
27. Wang, Z., Wu, Z., Agarwal, D., Sun, J.: Medclip: contrastive learning from unpaired medical images and text. In: Empirical Methods in Natural Language Processing (EMNLP), pp. 1–12 (2022)
28. Wu, C., Zhang, X., Zhang, Y., Wang, Y., Xie, W.: Medklip: medical knowledge enhanced language-image pre-training for x-ray diagnosis. In: ICCV (2023)
29. Zanella, M., Gérin, B., Ayed, I.B.: Boosting vision-language models with transduction. arXiv preprint arXiv:2406.01837 (2024)
30. Zhang, Y., Jiang, H., Miura, Y., Manning, C.D., Langlotz, C.P.: Contrastive learning of medical visual representations from paired images and text. In: MHLC (2022)
31. Ziko, I., Dolz, J., Granger, E., Ayed, I.B.: Laplacian regularized few-shot learning. In: International Conference on Machine Learning, pp. 11660–11670. PMLR (2020)

MAGDA: Multi-agent Guideline-Driven Diagnostic Assistance

David Bani-Harouni[1,2(✉)], Nassir Navab[1,2], and Matthias Keicher[1,2]

[1] Computer Aided Medical Procedures, School of Computation, Information and Technology, Technical University of Munich, Garching, Germany
[2] Munich Center for Machine Learning (MCML), Munich, Germany
david.bani-harouni@tum.de

Abstract. In emergency departments, rural hospitals, or clinics in less developed regions, clinicians often lack fast image analysis by trained radiologists, which can have a detrimental effect on patients healthcare. Large Language Models (LLMs) have the potential to alleviate some pressure from these clinicians by providing insights that can help them in their decision-making. While these LLMs achieve high test results on medical exams showcasing their great theoretical medical knowledge, they tend not to follow medical guidelines. In this work, we introduce a new approach for zero-shot guideline-driven decision support. We model a system of multiple LLM agents augmented with a contrastive vision-language model that collaborate to reach a patient diagnosis. After providing the agents with simple diagnostic guidelines, they will synthesize prompts and screen the image for findings following these guidelines. Finally, they provide understandable chain-of-thought reasoning for their diagnosis, which is then self-refined to consider inter-dependencies between diseases. As our method is zero-shot, it is adaptable to settings with rare diseases, where training data is limited, but expert-crafted disease descriptions are available. We evaluate our method on two chest X-ray datasets, CheXpert and ChestX-ray 14 Longtail, showcasing performance improvement over existing zero-shot methods and generalizability to rare diseases.

Keywords: Clinical guidelines · Large Language Models · Zero-shot classification

1 Introduction

Radiology holds a critical position in contemporary healthcare, being integral to the treatment and management of most patients. However, the healthcare sector is currently grappling with what has been termed the "radiologist shortage" [12].

Supplementary Information The online version contains supplementary material available at https://doi.org/10.1007/978-3-031-73471-7_17.

In the UK, this shortage stands at 29% and is predicted to worsen, reaching 40% within the next four years [15]. This effect is exacerbated in rural hospitals or clinics in less developed regions of the world, where the population per radiologist is much greater [11,18]. When there is a lack of radiologists, the clinicians with patient contact have to either miss out on valuable radiological information or evaluate that information themselves without proper training. Large Language Models (LLMs) have recently demonstrated remarkable potential for reasoning and solving complex problems, presenting an opportunity to address this challenge [14]. However, in a clinical context, deterministic models strictly adhering to evidence-based medical guidelines are preferred over creative but unpredictable LLM outputs. Moreover, generalist LLMs like GPT-4 [1] may lack domain-specific knowledge required for accurate diagnosis or have outdated medical insights. Consequently, providing LLMs access to relevant clinical knowledge sources, such as guidelines encapsulating the medical community's consensus, is critical for effective diagnostic processes, particularly for rare diseases with limited data. In contrast to visual instruct tuning [7], prompting Vision-Language Models (VLMs) is an intriguing approach to enabling LLMs to understand the content of images without the need to retrain. Recent explorations have successfully employed contrastive language-image pertaining (CLIP) [10] for few- and zero-shot classification of common diseases in chest X-rays [2,13,16,20,22,23]. Building on this, Xplainer [9] introduced a classification-by-description approach, querying a vision-language model for image observations indicative of a disease, providing inherent explainability. However, it naively averages concept probabilities, failing to account for dependencies between these concepts. To address those limitations, we propose MAGDA (Multi-Agent Guideline-driven Diagnostic Assistance), a multi-agent framework that unifies the incorporation of clinical guidelines as knowledge sources, dynamic prompting of a vision-language model for LLM understanding of radiology images, and a transparent diagnosis reasoning following the domain-specific knowledge provided by clinical guidelines. We show that this approach achieves state-of-the-art performance on zero-shot classification tasks of pathologies in the CheXpert dataset [5] and rare diseases in the ChestXRay 14 Longtail dataset [4]. Our key contributions are:

- An end-to-end guideline-driven approach that requires only a clinical guideline and a medical image as input to perform zero-shot diagnosis
- Novel dynamic prompting of vision-language models to enable LLMs to screen medical images for unseen diseases without the need for fine-tuning
- A transparent reasoning process through chain-of-thought reasoning, providing insights into the diagnostic decision-making

2   Methodology

2.1   Model Overview

We propose MAGDA, a multi-agent zero-shot method that can work with expert-crafted disease descriptions to provide transparent decision support. A general

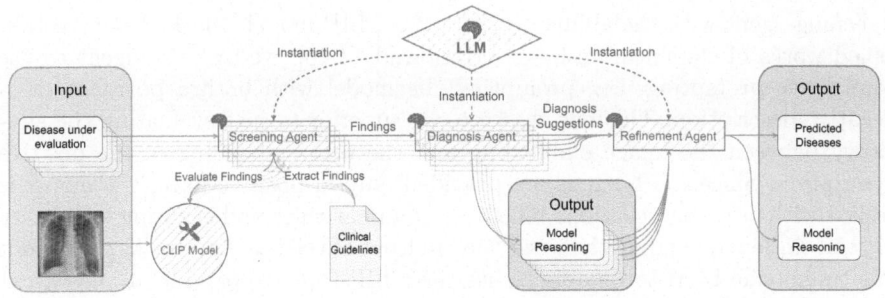

Fig. 1. Schematic overview of the proposed method MAGDA.

overview of the method is shown in Fig. 1. The LLMs used are not fine-tuned and all adaptions to the tasks are performed in-context. The multi-agent system consists of three agents that take over different tasks in the diagnosis procedure:

1. **Screening agent** $\mathcal{S}$: This agent handles the image analysis. As an Augmented Language Model [8] with tool-using capabilities, it can prompt a CLIP model [10] to evaluate fine-grained image findings according to given diagnosis guidelines.
2. **Diagnosis agent** $\mathcal{D}$: This agent is given the image findings from the screening agent and is tasked to reason about these findings to reach a diagnosis for the patient
3. **Refinement agent** $\mathcal{R}$: This agent is responsible for refining the predictions of the diagnosis agent by considering inter-dependencies of diseases and evaluating the quality of the reasoning. It then gives the final diagnosis prediction for the patient at hand.

2.2 Screening Agent

When diagnosing a specific patient $p \in P$, this agent is run once for every disease $d \in D$ that we want to evaluate. It is presented with expert-crafted fine-grained image findings and returns the positive or negative findings present in the image following the given diagnosis guidelines.

$$\mathcal{S}_d^p(G_d, d) \rightarrow F_d^p,$$

where G_d are the disease guidelines, d is the condition under evaluation, and F_d^p are the patient and disease-specific positive and negative image findings. The guidelines G_d can be provided in either an already expert-crafted fine-grained list of disease-specific image findings or an unstructured disease description from which the model can extract these fine-grained image findings. For example, in the case of an enlarged cardiomediastinum, one image finding may be described as "Abnormal contour of the heart border". In order to screen the image for the presence or absence of these image findings, we augmented the screening agent with the ability to prompt a CLIP model [10]. Following established works of classification-by-description [9,13,16], we task the agent to use contrastive prompting, i.e., prompting the model with both a positive and a negative description. This has shown to be superior to just evaluating the similarity between the image embedding and the text embedding of the positive description. Since the findings are provided only in positive form, the model is tasked with creating negations following grammar rules and ensuring sensibility beyond simply appending the word "no" in front of the description. As we are not fine-tuning the LLM to be able to use the CLIP-tool, we provide an in-context description of the tool together with instructions on how to use it. Through one example, it is further tasked with following the report style template from Xplainer [9]. During inference, the tool can be called using the call:

CLIP: [positive description] / [negative description] ->

Once the descriptions have been extracted from the model output, we provide them to the CLIP model. Specifically, we employ the image and text encoder from BioVil-T [2]. After computing the cosine similarity of the image embedding with the positive and negative text embeddings, we calculate the softmax over these two similarities to get the final probabilities of each finding. Initial results on the validation set showed that the BioVil-T model tends to over-predict positive findings, so we only count a positive finding if its probability exceeds a threshold ψ. Finally, depending on whether the tool returns a positive or negative result for the given description, we append "Positive" or "Negative" to the inference text and continue the LLM inference from there. After all given descriptions have been evaluated, the collected positive and negative disease descriptions are passed to the diagnosis agent.

2.3 Diagnosis Agent

The diagnosis agent is again run once per patient and condition under evaluation. It is given the list of findings extracted from the image by the screening agent and returns a positive or negative prediction, including the reasoning for that decision.

$$\mathcal{D}_d^p(F_d^p, d) \rightarrow p_d^p, r_d^p,$$

where p_d^p is the binary disease prediction for patient p and disease d, and r_d^p is the reasoning for that prediction. It has been shown that LLM reasoning can be significantly improved by chain-of-thought prompting [21], a prompt engineering technique where the model is asked to provide step-by-step reasoning before answering a question. Additionally, to an increase in reasoning capabilities, this reasoning makes the method inherently explainable. As the model provides explanations for its predictions, clinicians can use these explanations to evaluate the decision process and increase trust in the model output. Specifically, we ask the model to provide reasoning before answering the question "Does the patient have [d]?". We prompt the model to use a specified format to make parsing the model output possible, ending the reasoning process with the sentence:

"Therefore, my answer is: [yes/no]." Once the various predictions and reasonings have been collected, they are passed to the refinement agent for the final patient diagnosis.

2.4 Refinement Agent

The refinement agent is run once per patient. It is presented with all positive disease predictions and the diagnosis agent's reasonings for these predictions. It returns the final patient prediction.

$$\mathcal{R}^p(\{(p_d^p, r_d^p)|d \in D, p_d^p \text{ is positive}\}) \rightarrow \{\hat{p}_d^p|d \in D\}$$

The refinement agent is tasked with evaluating the provided reasoning. So far, every disease has been evaluated on its own in order to not overload the agents. At this step in the diagnosis process, inter-dependencies between diseases can be considered. The refinement agent is queried for every disease under evaluation if that disease is present or not and again asked to provide chain-of-thought reasoning for that decision. From the model replies we parse the final patient predictions $\hat{p}^p$. "No Finding" is predicted if all other disease predictions are negative.

3 Experimental Setup

Datasets and Evaluation Metrics. We evaluate our method on two chest X-ray datasets, CheXpert [5] and ChestXRay 14 Longtail [4,19]. The CheXpert dataset includes manually annotated validation and test sets comprising 200 and 500 patients, respectively. It encompasses 14 different categories, featuring "No Finding", 12 pathology labels, e.g., "Pneumonia", and a class "Support Devices". On CheXpert, we perform multi-label classification. Most comparable methods evaluate using the Area Under the ROC-curve (AUC) metric. As our method generates discrete predictions, threshold-independent metrics, like AUC, cannot sensibly be evaluated. Instead, we report micro and macro F1-score, precision, and recall. The CLIP finding probability threshold ψ, which is used to combat the over-prediction of the CLIP model, is set to 0.55 based on experiments on the validation set.

The ChestXRay14 Longtail dataset is an extension of the common ChestXRay 14 dataset by adding five additional disease findings, expanding the classification to 20 categories. These are divided into 7 head classes (most common), 10 medium classes (moderately common), and 3 tail classes (least common). The dataset includes a balanced validation and test set, each offering 15 or 30 images per class, respectively, to ensure comprehensive coverage and evaluation capabilities across the spectrum of conditions. We evaluate on this balanced test set with equal number of cases per class. Here, we perform single-label classification and report the accuracy on the three tail classes. In this setting, we prompt the CLIP model without description negation. As the screening and diagnosis agents always perform multi-label classification, we further adapt our refinement agent to decide on exactly one positive prediction.

Implementation Details. The backbone of our method lies in a powerful LLM instantiated in different ways as the various agents. Unless stated otherwise, we employ the Mixtral $8 \times 7B$ instruct model from Mistral AI [6]. This model provides a good trade-off between memory efficiency, inference speed, and model capabilities. We use a 4-bit GPTQ quantized version of the model [3]. This reduces the memory requirements while keeping the loss in model accuracy minimal. Thus processed we are able to run text generations on a single Nvidia A40 GPU using a temperature of 0.8. Where available, we used the image findings from the public Xplainer repository [9] as guidelines to ensure fair comparison. Where not available, we used a similar approach to Xplainer of prompting GPT-4 to generate candidate findings, and correcting them based on text book knowledge.

4   Results and Discussion

Table 1. Test set results for zero-shot classification on the CheXpert dataset. MAGDA (nG) is our proposed method without the use of guidelines, relying on LLM knowledge.

Method	F1-score		Precision		Recall	
	micro	macro	micro	macro	micro	macro
CheXzero	35.69	33.50	25.58	**37.72**	58.98	64.88
Xplainer	45.33	39.27	31.36	33.74	81.74	83.27
MAGDA (nG)	42.94	36.50	29.39	30.82	79.62	82.07
MAGDA	**46.18**	**39.58**	**31.93**	33.43	**83.43**	**83.47**

Table 2. Test set results on the ChestXray 14 Longtail dataset. Accuracy is the classification accuracy on the rare tail classes. Methods above the line are fully supervised.

Method	Zero-shot	Accuracy
ResNet-50 [4]	×	1.7
Decoupling-cRT [4]	×	30.0
CheXzero	✓	12.3
Xplainer	✓	8.2
MAGDA	✓	**18.5**

In Table 1, we compare with state-of-the-art zero-shot classification methods CheXzero [16] and Xplainer [9] on the CheXpert test dataset. Because CheXzero is only evaluated on the six competition pathologies in the original paper, we

Table 3. Comparison of zero-shot classification of the diagnosis agent, i.e., before refinement with the refinement agent, using different LLMs as a reasoning backbone on the test set of the CheXpert dataset.

Reasoning Model	F1-score		Precision		Recall	
	micro	macro	micro	macro	micro	macro
GPT-4	**46.87**	**41.10**	**32.17**	**34.17**	86.31	85.13
Llama2 70B chat	46.36	40.72	31.62	33.48	**86.87**	**88.01**
Mixtral 8 × 7B instruct	45.31	39.70	30.85	33.41	85.25	86.62

use their public code and model to evaluate on all CheXpert classes. Most state-of-the-art methods only report the AUC, we can therefore only compare with methods with published code and calculate their respective F1-score, precision, and recall. For example, a comparison with Seibold et al. [13] or ELIXR [24] was not possible for that reason. We outperform all comparable methods on zero-shot classification on all metrics except macro precision, where CheXzero has a better score at the expense of a much lower recall. Here, we also compare the guideline-driven approach with the generation of findings by the model itself, showing that the provision of guidelines to our method increases performance. Table 2 shows a comparison on the ChestXray14 Longtail dataset with the same zero-shot methods as before and additionally with two fully supervised methods trained on that dataset [4]. Here, we again outperform both zero-shot methods on the tail classes. Notably, we even reach a higher accuracy than a simple supervised method trained on the 68,058 training samples. Only a highly tuned method employing decoupled training [4] achieves a higher accuracy. These results show that the provision of detailed guidelines describing the diagnosis of rare and lesser-known diseases can help with diagnostic accuracy. In Table 3, we compare exchanging our Mixtral 8 × 7B instruct model for other well-known LLMs, namely Llama 2 70B chat [17] and GPT-4 [1]. While both alternative models reach higher performance, this comes at the cost of higher computational needs in the case of Llama2 70B chat, or dependence on a proprietary API.

Ablation Studies. We now want to look at the benefits of different aspects of our method. First, in Table 4, we compare the rule-based negation by simply appending a "no" before the finding description with the LLM-created finding negation done by the screening agent. We see that the latter results in better performance, highlighting the benefit of descriptions that are more aligned with natural language and thus the style of radiology reports. In Table 5, different approaches for the refinement agent are compared. We compare combinations of chain-of-thought reasoning and including the CheXpert disease graph in textual form [5]. This disease graph models the dependencies between classes, e.g. "Enlarged Cardiomediastinum" being a sign of "Cardiomegaly". However, both in the case of using chain-of-thought reasoning and not using it, the inclusion of the disease graph decreases the performance.

Table 4. Comparison of naive finding negation with the improved negation by the Screening agent. Results are reported before the refinement by the refinement agent on the CheXpert validation set.

CLIP prompting	F1-score	
	micro	macro
Naive negation	46.79	41.75
LLM negation	**48.00**	**42.10**

Table 5. Comparison of different refinement approaches on the CheXpert validation set. CoT = chain-of-thought reasoning, DG = disease graph.

CoT	DG	F1-score	
		micro	macro
×	×	48.10	41.56
×	✓	47.79	40.39
✓	×	**49.17**	**42.05**
✓	✓	47.52	40.70

Qualitative Results. In Fig. 2, we show a qualitative example of the reasoning provided by the diagnosis agent. The agent is presented with conflicting findings regarding the diagnosis of an enlarged cardiomediastinum. Instead of naively aggregating them, giving each finding the same importance, the diagnosis agent can differentiate between more and less relevant findings and can employ reason to reach a diagnosis. This example also shows the high dependence on the correctness of the image findings identified by the CLIP model. Even perfect reasoning cannot reach the correct conclusion if it works with wrong information. Therefore, improvements in the performance of VLMs can directly translate into further improvement of our method.

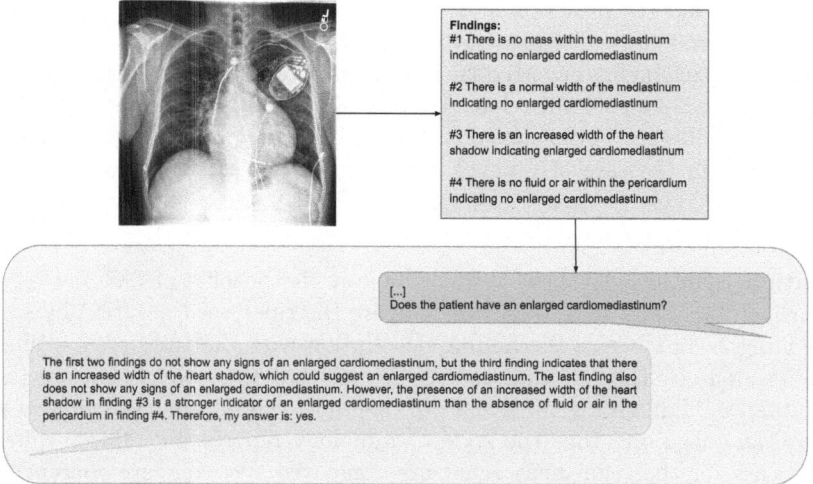

Fig. 2. A qualitative example of the model reasoning.

5 Conclusion

In this paper, we presented MAGDA, a novel multi-agent approach that integrates clinical guidelines, dynamic vision-language model prompting, and large language model reasoning to address the challenges of diagnostic assistance using LLMs. Our approach leverages the strengths of both LLMs and VLMs, enabling the zero-shot classification of diseases without the need for model retraining or fine-tuning. This guideline-driven methodology not only facilitates accurate diagnoses from medical images but also introduces a transparent reasoning process that enhances the explainability and trustworthiness of the diagnostic outcomes. Our evaluation on the CheXpert and ChestXray14 LT datasets demonstrates the effectiveness of our approach, particularly in scenarios involving rare diseases where traditional diagnostic methods are often hindered by data scarcity. By incorporating domain-specific knowledge through clinical guidelines and employing dynamic prompting techniques, we improve diagnostic accuracy and model trustworthiness.

Acknowledgement. The authors gratefully acknowledge the financial support by the Bavarian Ministry of Economic Affairs, Regional Development and Energy (StMWi) under project ThoraXAI (DIK-2302-0002).

References

1. Achiam, J., et al.: Gpt-4 technical report. arXiv preprint arXiv:2303.08774 (2023)
2. Bannur, S., et al.: Learning to exploit temporal structure for biomedical vision-language processing. In: Proceedings of the IEEE/CVF Conference on Computer Vision and Pattern Recognition, pp. 15016–15027 (2023)
3. Frantar, E., Ashkboos, S., Hoefler, T., Alistarh, D.: Gptq: Accurate post-training quantization for generative pre-trained transformers. arXiv preprint arXiv:2210.17323 (2022)
4. Holste, G., et al.: Long-Tailed classification of thorax diseases on chest X-Ray: a new benchmark study. In: Nguyen, H.V., Huang, S.X., Xue, Y. (eds.) Data Augmentation, Labelling, and Imperfections. DALI 2022. Lecture Notes in Computer Science, vol. 13567, pp. 22–32. Springer, Cham (2022). https://doi.org/10.1007/978-3-031-17027-0_3
5. Irvin, J., et al.: Chexpert: a large chest radiograph dataset with uncertainty labels and expert comparison. In: Proceedings of the AAAI Conference on Artificial Intelligence, vol. 33, pp. 590–597 (2019)
6. Jiang, A.Q., et al.: Mixtral of experts. arXiv preprint arXiv:2401.04088 (2024)
7. Liu, H., Li, C., Wu, Q., Lee, Y.J.: Visual instruction tuning. In: Oh, A., Neumann, T., Globerson, A., Saenko, K., Hardt, M., Levine, S. (eds.) Advances in Neural Information Processing Systems. vol. 36, pp. 34892–34916. Curran Associates, Inc. (2023)
8. Mialon, G., et al.: Augmented language models: a survey. arXiv preprint arXiv:2302.07842 (2023)
9. Pellegrini, C., Keicher, M., Özsoy, E., Jiraskova, P., Braren, R., Navab, N.: Xplainer: From x-ray observations to explainable zero-shot diagnosis. arXiv preprint arXiv:2303.13391 (2023)

10. Radford, A., et al.: Learning transferable visual models from natural language supervision. In: International Conference on Machine Learning, pp. 8748–8763. PMLR (2021)
11. Ramli, N.M., Zain, N.R.M.: The growing problem of radiologist shortage: Malaysia's perspective. Korean J. Radiol. **24**(10), 936 (2023)
12. Rimmer, A.: Radiologist shortage leaves patient care at risk, warns royal college. BMJ: Br. Med. J. (Online) **359** (2017)
13. Seibold, C., Reiß, S., Sarfraz, M.S., Stiefelhagen, R., Kleesiek, J.: Breaking with fixed set pathology recognition through report-guided contrastive training. In: Wang, L., Dou, Q., Fletcher, P.T., Speidel, S., Li, S. (eds) Medical Image Computing and Computer Assisted Intervention – MICCAI 2022. MICCAI 2022. Lecture Notes in Computer Science, vol. 13435, pp. 690–700 (2022). Springer, Cham. https://doi.org/10.1007/978-3-031-16443-9_66
14. Singhal, K., Azizi, S., Tu, T., Mahdavi, S.S., Wei, J., Chung, H.W., Scales, N., Tanwani, A., Cole-Lewis, H., Pfohl, S., et al.: Large language models encode clinical knowledge. Nature **620**(7972), 172–180 (2023)
15. The Royal College of Radiologists: Clinical radiology census report 2022. The Royal College of Radiologists (Online) (2022)
16. Tiu, E., Talius, E., Patel, P., Langlotz, C.P., Ng, A.Y., Rajpurkar, P.: Expert-level detection of pathologies from unannotated chest x-ray images via self-supervised learning. Nat. Biomed. Eng. **6**(12), 1399–1406 (2022)
17. Touvron, H., et al.: Llama 2: Open foundation and fine-tuned chat models. arXiv preprint arXiv:2307.09288 (2023)
18. Vu, L.D., Nguyen, H.T.T., Nguyen, T.N., Pham, T.M.: The growing problem of radiologist shortage: Vietnam's perspectives. Korean J. Radiol. **24**(11), 1054 (2023)
19. Wang, X., Peng, Y., Lu, L., Lu, Z., Bagheri, M., Summers, R.M.: Chestx-ray8: hospital-scale chest x-ray database and benchmarks on weakly-supervised classification and localization of common thorax diseases. In: Proceedings of the IEEE Conference on Computer Vision and Pattern Recognition, pp. 2097–2106 (2017)
20. Wang, Z., Wu, Z., Agarwal, D., Sun, J.: Medclip: Contrastive learning from unpaired medical images and text. arXiv preprint arXiv:2210.10163 (2022)
21. Wei, J., et al.: Chain-of-thought prompting elicits reasoning in large language models. Adv. Neural. Inf. Process. Syst. **35**, 24824–24837 (2022)
22. Windsor, R., Jamaludin, A., Kadir, T., Zisserman, A.: Vision-language modelling for radiological imaging and reports in the low data regime. In: Medical Imaging with Deep Learning (2023)
23. Wu, C., Zhang, X., Zhang, Y., Wang, Y., Xie, W.: Medklip: medical knowledge enhanced language-image pre-training for x-ray diagnosis. In: Proceedings of the IEEE/CVF International Conference on Computer Vision, pp. 21372–21383 (2023)
24. Xu, S., Yang, L., Kelly, C., Sieniek, M., Kohlberger, T., Ma, M., Weng, W.H., Kiraly, A., Kazemzadeh, S., Melamed, Z., et al.: Elixr: Towards a general purpose x-ray artificial intelligence system through alignment of large language models and radiology vision encoders. arXiv preprint arXiv:2308.01317 (2023)

Author Index

Z. Deng et al. (Eds.): MedAGI 2024, LNCS 15184, pp. 173–174, 2025.
https://doi.org/10.1007/978-3-031-73471-7